AF775105

Topographical and Pathotopographical Medical Atlas of the Pelvis, Spine, and Limbs

Scrivener Publishing
100 Cummings Center, Suite 541J
Beverly, MA 01915-6106

Publishers at Scrivener
Martin Scrivener (martin@scrivenerpublishing.com)
Phillip Carmical (pcarmical@scrivenerpublishing.com)

Topographical and Pathotopographical Medical Atlas of the Pelvis, Spine, and Limbs

Z. M. Seagal

Scrivener Publishing

WILEY

This edition first published 2019 by John Wiley & Sons, Inc., 111 River Street, Hoboken, NJ 07030, USA and Scrivener Publishing LLC, 100 Cummings Center, Suite 541J, Beverly, MA 01915, USA
© 2019 Scrivener Publishing LLC
For more information about Scrivener publications please visit www.scrivenerpublishing.com.

All rights reserved. No part of this publication may be reproduced, stored in a retrieval system, or transmitted, in any form or by any means, electronic, mechanical, photocopying, recording, or otherwise, except as permitted by law. Advice on how to obtain permission to reuse material from this title is available at http://www.wiley.com/go/permissions.

Wiley Global Headquarters
111 River Street, Hoboken, NJ 07030, USA

For details of our global editorial offices, customer services, and more information about Wiley products visit us at www.wiley.com.

Limit of Liability/Disclaimer of Warranty
While the publisher and authors have used their best efforts in preparing this work, they make no representations or warranties with respect to the accuracy or completeness of the contents of this work and specifically disclaim all warranties, including without limitation any implied warranties of merchantability or fitness for a particular purpose. No warranty may be created or extended by sales representatives, written sales materials, or promotional statements for this work. The fact that an organization, website, or product is referred to in this work as a citation and/or potential source of further information does not mean that the publisher and authors endorse the information or services the organization, website, or product may provide or recommendations it may make. This work is sold with the understanding that the publisher is not engaged in rendering professional services. The advice and strategies contained herein may not be suitable for your situation. You should consult with a specialist where appropriate. Neither the publisher nor authors shall be liable for any loss of profit or any other commercial damages, including but not limited to special, incidental, consequential, or other damages. Further, readers should be aware that websites listed in this work may have changed or disappeared between when this work was written and when it is read.

Library of Congress Cataloging-in-Publication Data

ISBN 978-1-11961-4-258

Cover image: Courtesy of Z. M. Zeagal
Cover design by Kris Hackerott

Set in size of 13pt and Minion Pro by Exeter Premedia Services Private Ltd., Chennai, India

10 9 8 7 6 5 4 3 2 1

Contents

Introduction

This color atlas of topographic and pathotopographic human anatomy is a fundamentally and practically important book designed for doctors of all specializations and students of medical schools. Here you can find almost everything that is connected with the topographic and pathotopographic human anatomy, including original graphs of logical structures of topographic anatomy and development of congenital abnormalities, topography of different areas in layers, pathotopography, and computer and magnetic resonance imaging (MRI) of topographic and pathotopographic anatomy. You can also find new theoretical and practical sections of topographic anatomy developed by the author himself, which are published for the first time. They are practically important for mastering the technique of operative interventions and denying the possibility of iatrogenic complications during operations.

All the pictures are completely original and are drawn and colored at a highly professional level. There are also illustrations of individual and age differences of human organs in the chapter of variant topographical anatomy. In the chapter of topographical anatomy, you can find holothopy, sceletothopy, synthopy of organs and main anatomical formations, facial vaginae, reticular spaces, branches of nerves and vessels, collateral vessels, and zones of sensory and motoric innervation of nerve trunci. Pathotopographic anatomy is displayed by the examples of typical pathology of different areas.

One of the distinguishing features of this book is that topographic and anatomic structures are presented both in corpses and postoperative patients. This data is gathered by the use of MRI and transillumination of organs and tissues.

Unfortunately, nowadays there is no publication of modern atlases of clinical, topographic, and pathotopographic anatomy due to several reasons. Firstly, this subject is separated only in Russia and Hungary. In other countries, it is either absent or exists as a part of similar subjects like anatomy or surgery. Some of the themes presented in this book like pathotopographic, variant, normal, and topographic anatomy, trans-illumination, pathotopographic and topographic anatomy of patients and healthy people are original studies of the author of this book.

This color atlas can serve as a textbook for routine necessities of theoretical and practical work of physiologists and pathophysiologists, anatomists and topographic anatomists, pathoanatomists, therapists, surgeons, gynecologists and obstetricians, neurologists and dermatovenerologists, radiologists, reanimatologists, anesthesiologists, otorhinolaryngologic, traumatologists, orthopedists, pediatricians, dentists, and other doctors of narrow specializations like angiosurgeons, urologists, plastic surgeons, neurosurgeons, etc.

Topographic and pathotopographic anatomy presented in this book can be divided into surgical anatomy, therapeutic anatomy, stomatological anatomy, pediatric anatomy, etc. There are also some specific and non-specific indications of pathotopography and variants of physiology, differences between physiology and pathology, and ways of using ultrasound topographic and pathotopographic anatomy. It should be pointed out that studying a living human organism in physiological and pathological conditions is the most effective way for reliable diagnostics. Here you can also find descriptions of less traumatic and non-traumatic operations, as well as effective ways of treatment.

Topographic clinical anatomy changes in the course of a disease and recovery, iatrogenic manipulations, as well as therapeutic and surgical interventions. Some of these situations are described in this book. The author of this book will eagerly accept all constructive criticism and take it into consideration in the next editions of the book.

Part 1: The Pelvis

Topographic Anatomy of the Pelvis

Limits. The pelvis is a part of the body located between the abdomen and lower limbs. It is limited outside by the pelvic bones, sacrum, coccyx and perineum at the bottom.

External reference points: the iliac crests (*crista iliaca),* anterior superior iliac spines (spina iliaca anterior superior), pubic tubercle (tuberculum pubicum), pubic symphysis, (symphysis pubica), dorsal surface of the sacrum (facies dorsalis os sacrum), tailbone (os coccyges); ischial tuberosity (tuber ischiadicum), the greater trochanter of the femur (trochanter major ossis femoralis), subpubicalis angle (angulus subpubicus).

Individual, Gender and Age Differences:

1. The individual features of the pelvis consist of different ratios of the longitudinal and transverse diameters of the pelvic ring. At one extreme type of variability, the longitudinal diameter is greater than the transverse one, the pelvis is "squeezed" from the sides, and the axes of the pelvic organs are often inclined to the sacrum. With a different type of variability, the longitudinal diameter is less than the transverse one, the pelvis is "compressed" in

the anteroposterior direction, and the axes of the pelvic organs are inclined toward the pubic symphysis.

2. The gender differences. The wings of the iliac bones in women are more horizontal, therefore in women the pelvis is wider and lower than that of men. The inferior branches of pubic bones in women are located at an obtuse angle and form a pubic arch (arcus pubis); in men they are located at an acute angle and form a subpubic angle (angulus subpubicus). The pelvic cavity in women has the shape of a curved cylinder; in men it has the shape of a curved cone. The incline of the pelvis (inclination pelvis) is the angle between the horizontal plane and the plane of the superior aperture of the pelvis. In women it's 55–60°, in men – 50–55°.
3. Age differences. By the time of birth, the pelvic consists of three parts, each of which has ossification centers. All three parts – iliac, ischial and pubic – are connected inter-layer cartilage in the area of the acetabulum. The iliac fossa is almost absent. By the age of seven, the size of the pelvic bone doubles, then the growth slows to 12 years. By the age of 13–18, parts of the pelvic bone coalesce, and their ossification ends. The synostosis of all elements of the hip-bone is completed by the age of 25.

The walls of the pelvis. The pelvis is restricted to the pubic, iliac, ischium, sacrum and coccyx.

The pelvic floor is formed by the diaphragm of the pelvis (diaphragma pelvi) and partly by the urogenital diaphragm (diaphragma urogenitale).

The genitourinary diaphragm, diaphragma urogenitale (Figures 1–5) is formed by two muscles: the deep perineal transverse muscle and the external sphincter of the urethra.

The diaphragm of the pelvis is formed by two muscles: a muscle lifting the anus (m. levator ani) and the coccygeal muscle (m. soccygeus).

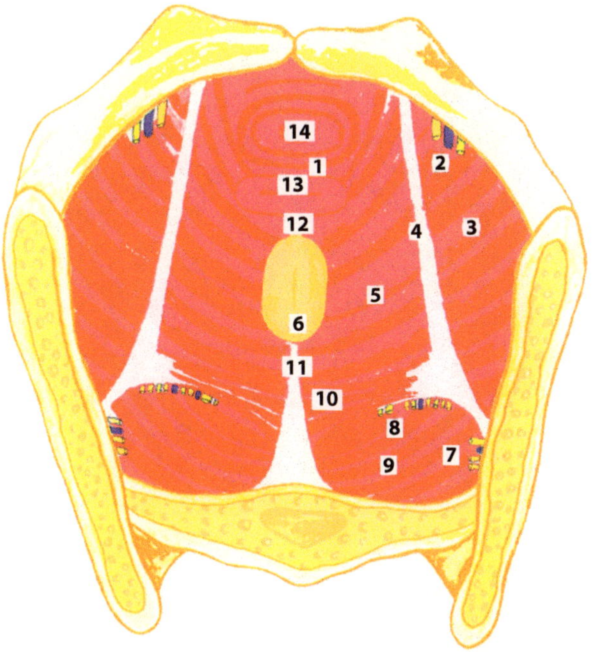

Figure 1 Urogenital diaphragm in women.
1 – diaphragma urogenitalis; 2 – nn. et vasa obturatorii; 3 – m. obturatorius internus; 4 – arcus tendineus m. levatoris ani; 5 – m. levator ani; 6 – anus; 7 – nn. et vasa glutei superiores; 8 – n. ischiadicus et fasc. gluteus inferior; 9 – m. piriformis; 10 – m. coccygeus; 11 – lig. anococcygeus; 12 – centrum tendineum perinei; 13 – vagina; 14 – urethra

The floors of the cavity of the small pelvis. The cavity of the small pelvis is divided into three floors: upper one – peritoneum (cavum pelvis peritoneale), middle one – subperitoneal (cavum pelvis subperitoneale) and lower one – subcutaneous or perineal (cavum pelvis subcutaneum s. perineale).

Peritoneum in the male pelvis. In the male pelvis, the peritoneum from the anterior wall of the abdomen passes to the bladder, covering the upper wall and partly the lateral and posterior ones. On the sides of the bladder the peritoneum forms fossae paravesicales. At the symphysis level, the peritoneum forms plica vesicalis transversa. Going down the back wall of the bladder, the peritoneum covers the medial edges of the vas deferens and the apex of the seminal glands, and passes to the rectum, forming the recto-vesical pouch (excavatio rectovesicalis).

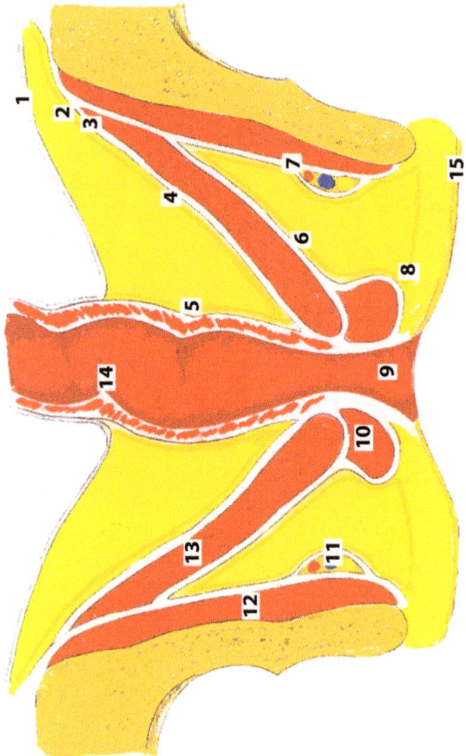

Figure 2 The front section of the pelvis.
1 – peritoneum; 2 – fascia intraperitonealis; 3 – arcus tendneus; 4 – the parietal peritoneum; 5 – the visceral peritoneum; 6 – the deep perineal fascia; 7 – fascia m. obturatorii; 8 – the superficial perineal fascia; 9 – anus; 10 – m. sphincter ani externi; 11 – canalis pro vasa pudendi et n.; 12 – m. obduratorius internus; 13 – m. levator ani; 14 – rectum; 15 – derma perinei

The ampulla is covered on three sides, and the lower part of the rectum is not covered with the peritoneum at all.

Peritoneum in the female pelvis. In the female pelvis, the peritoneum also passes from the anterior abdominal wall to the bladder, forming a transverse fold, and then covers its superior and posterior walls. After this, it passes to the anterior surface of the uterus at the level of the isthmus, where it forms a shallow vesicouterine pouch (excavatio vesicouterina). On the anterior surface of the uterus, the peritoneum covers only the body of the uterus. On the posterior surface of the uterus, the peritoneum covers the body of the uterus, the

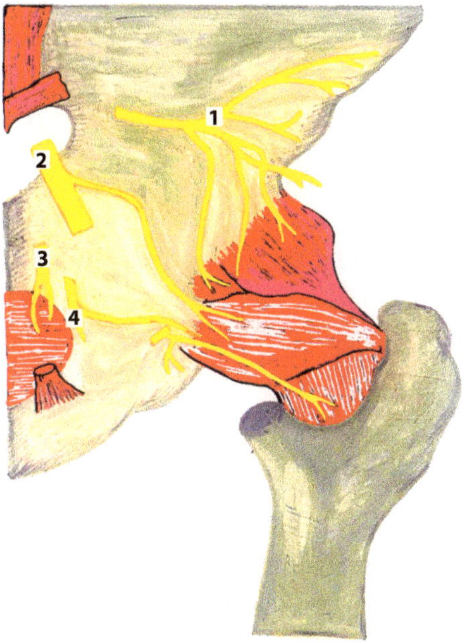

Figure 3 The innervation of the hip joint.
1 – n. gluteus superior; 2 – n. ischiadicus; 3 – n. pudendus; 4 – n. gluteus inferior

supramarginal part of the cervix and the posterior vaginal vault and passes to the rectum, forming a deep recto-uterine pounch (the excavatio rectouterine) or Douglas' pouch. It is limited by folds of the peritoneum – plicae rectouterinae, which extend to the anterior surface of the sacrum.

Fascia and cellular spaces, their connection with the cellular spaces adjacent regions. The parietal pelvic fascia (fascia pelvis parietalis) covers the anterior surface of the sacrum behind the back and is called the presacral fascia (fascia presacralis), on the sides it covers the parietal muscles (m. piriformis, m. obturatorius internus), the fascia of which have corresponding names (fascia obturatoria, fascia m. piriformis) in front – the posterior surface of the symphysis and the superior branches of the pubic bones, from below – the upper surface m. levator ani. The visceral pelvic fascia (fascia pelvis visceralis) lining the extraperitoneal parts of the pelvic organs in men and forms two sagittal branches extending from the pubic bones to the sacrum. Part of the visceral fascia that stretching from the pubic bone to the prostate

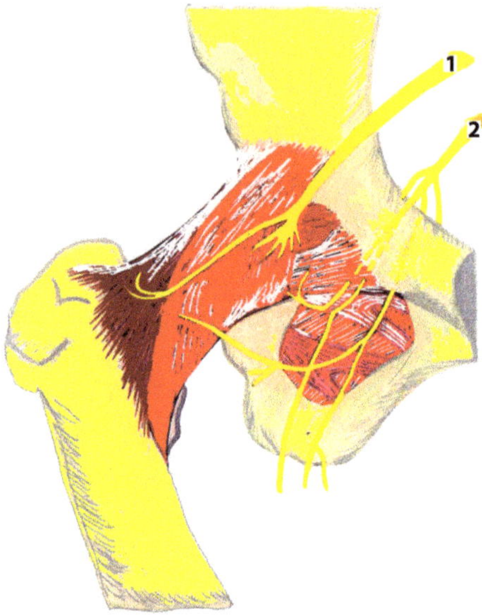

Figure 4 The innervation of the hip joint.
1 – n. femoralis; 2 – n. obturatorius

gland in men or to the bladder in women is called the puboprostatic (lig. puboprostaticum) or pubovesical (lig. pubovesicale) ligament. These ligaments are accompanied by bundles of smooth muscle fibers of m. puboprostaticus and m. pubovesicalis.

The sagittal branches of the visceral fascia that located behind the bladder also contain bundles of smooth muscle fibers that form several muscles: rectococcygeal (m. sacrococcygeus), rectovesical (m. recto-vesicalis) in men and rectouterine in women. Lining the pelvic organs, the visceral fascia of the pelvis forms a capsule of the rectum (Capsula Amussat) and a capsule of the prostate (Capsula Pirogov-Reitzia).

Cellular spaces. The major cellular spaces of the small pelvis are within the subperitoneal floor of the pelvis. Visceral cell spaces are the gaps between the organ's wall and the visceral fascia. There are the paravesical, paraprostatical, paravaginal, paracervical, pararectal visceral cellular spaces. There are four parietal cellular spaces in the subperitoneal floor of the small pelvis in the women: the retropubical (prevesical), two lateral and presacral (retrarectal). Men also have a fifth one, which is called the retrovesical cellular space.

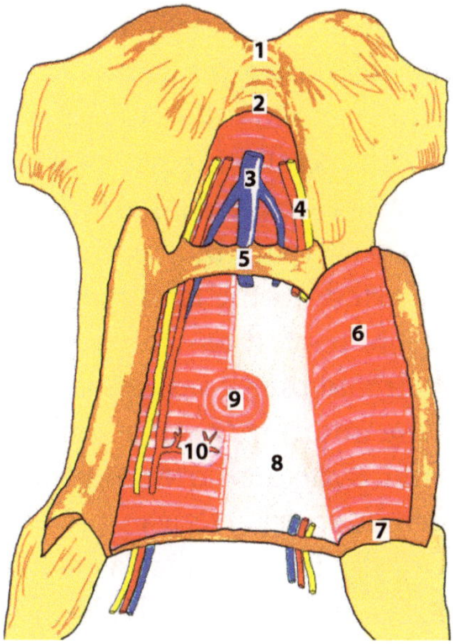

Figure 5 Urogenital diaphragm in men.
1 – symphisis ossium pubis; 2 – lig. arcuaturn pubis; 3 – v. penis; 4 – a. et n. penis; 5 – lig. transversum perinea; 6 – m. transversus perinei profundus; 7 – folium superficialis aponeurosis urogenitalis; 8 – folium profundum aponeurosis urogenitalis; 9 – urethra et m. sphincter urethrae; 10 – glandula bulbourethralis

The features in children. The fascia of the pelvis is very thin and friable. The parietal fascia is closely attached to the fascial vagina of the parietal vessels. The parietal and near-organ cellular spaces contain a small amount of adipose tissue, which makes the front and back Douglas spaces well expressed.

The internal iliac vessels. The main arterial trunk, supplying the organs of the small pelvis, is the internal iliac artery (a. iliaca interna). The branches of the internal iliac artery are:

- parietal: a. iliolumbalis, aa. sacrales laterales, a. obturatoria, a. glutea superior, a. glutea inferior;
- visceral: a. umbilicalis (a. vesicalis superior), a. vesicalis inferior, a. rectalis media, a. pudenda interna, a. uterina (a. ductus deferentis).

At the upper edge of the greater sciatic foramen a. iliaca interna is divided into the anterior and posterior trunks.

From the anterior trunk, mainly the visceral arteries originate: aa. uterina, vesicalis inferior, rectalis media; two parietal arteries, aa. umbilicalis et obturatoria, are directed anteriorly. The umbilical artery has two parts: the open part (pars patens) from which the superior vesicalis artery and the artery to the ductus deferens, and the closed one, pars occlusa. This obliterated part of the artery reaches the visceral fascia of the bladder and then goes along with it to the navel.

The ending part of the anterior trunk at infrapiriform foramen is divided into the internal pudendal artery (a. pudenda interna) and inferior gluteal (a. glutea inferior) arteries.

The posterior trunk of the internal iliac artery is divided into parietal arteries: aa. sacrales laterales, iliolumbalis and glutea superior. A. iliolumbalis with its lumbar and iliac branches anastomoses with the lumbar and intercostal arteries and with a. circumflexa ilium profunda and the gluteal arteries. Due to this, when there is occlusion of the common iliac artery, there is collateral blood flow.

Venous outflow from the pelvic organs is carried out first in the venous plexus, having the same names: pl. venosus rectalis, pl. venosus vesicalis, pl. venosus prostaticus, pl. venosus uterinus, pl. venosus vaginalis. Then the blood from these plexuses flows into the internal iliac vein, which is located deeper than the artery and medial from it. The parietal veins accompany the arteries in the form of paired vessels.

The children features. The arteries of the small pelvis of the newborn have their own particular qualities due to the features of the fetal blood supply: the common iliac, internal iliac (its anterior trunk) and umbilical arteries are represented by a single main vessel of the same diameter all along.

The sacral plexus (Figure 6). It is formed by the anterior branches of the IV and V lumbar and anterior branches of the I, II, III, IV sacral spinal nerves exiting through the anterior sacral foramen. It lies on the anterior surface of the piriformis muscle.

Short and long branches leave the sacral plexus. Short branches of n. obturatorius are guided along the side wall of the pelvis to the obturator

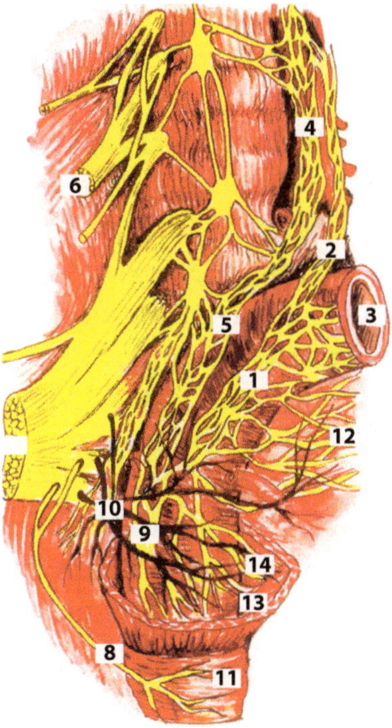

Figure 6 The Pelvic Plexus.
1 – plexus rectalis medius; 2 – plexus rectalis superior; 3 – rectum; 4 – plexus aorticus; 5 – plexus hypogasricus superior; 6 – plexus lumbalis; 7 – plexus sacralis; 8 – n. pudendus; 9 – plexus hypogasricus inferior; 10 – nn. pelvis interni; 11 – m. sphincter ani externus; 12 – plexus vesicalis; 13 – m. levator ani; 14 – plexus prostaticus

foramen. N. gluteus superior goes to the suprapiriform foramen with the same artery and vein. N. gluteus inferior and n. pudendus leave the cavity of the small pelvis through the infrapiriform foramen. In addition, n. pudendus, internal arteries and veins enter the ischioanal fossa through the lesser sciatic foramen.

Together with them, the long branches of the sacral plexus (n. ischiadicus и n.cutaneus femoris posterior) enter the gluteal region and are directed into the infrapiriform foramen with the inferior gluteal vessels.

The inferior hypogastric plexus (plexus hypogastricus inferior) is a vegetative plexus that includes post-nodal sympathetic branches, parasympathetic nodes, andsensitive fibers for the innervation of pelvic organs. It descends inthe form of a lamella from the sacrum to the bladder.

The lymphatic vessels and regional lymph nodes. Groups of lymph nodes: along the path of external and common iliac artery (from the free lower limb), along the path of the internal iliac artery (from the pelvic organs), retrorectal (from the sacrum and the coccyx). The outflow of lymph in the small pelvis takes place through three groups of nodes. The first is located along the internal iliac vessels (nodi iliaci interni). It collects lymph from the organs of the small pelvis. The second group (nodi iliaci externi et communis) is located along the external and common iliac arteries. It receives lymph from the lower limb, the lower parts of the abdominal wall, the superficial layers of the perineum, the external genitalia. The third group is the sacral nodes (nodi sacrales). They collect lymph from the posterior wall of the pelvis and the rectum. The nodes lying in the bifurcation of the common iliac arteries are called interiliac lymph nodes (nodi interiliaci). They receive lymph from both the pelvic organs and the lower limbs.

The Organs of the Male Pelvis

The topography of the bladder (Figures 7–11). The bladder is located in the anterior part of the small pelvis, behind the pubic bones and symphysis, the filled bladder in an adult goes beyond the pelvic cavity, rising above the pubic bones. It has the apex, body, fundus and neck. The wall of the bladder has well-expressed muscular and submucosal layers. In the region of the bottom of the bladder, the folds and submucosal layer are absent; the mucosa coalesces with the muscular membrane. Here a triangular area or Lieutaud triangle (trigonum vesicae urinariae.) is formed. The peritoneum passes from the anterior abdominal wall to the bladder and covers a very small part of the anterior wall, the superior and posterior walls of the bladder. It forms the transverse fold. Passing from the back wall to the rectum, the peritoneum forms the vesicorectal fold and recto-vesical pouch (excavatio rectovesicalis). In the subperitoneal region, the bladder has its own visceral fascia. The venous network of the bladder is located in a well-expressed layer of the fatty tissue between the wall of the bladder and the fascia in the the paravesical spaces.

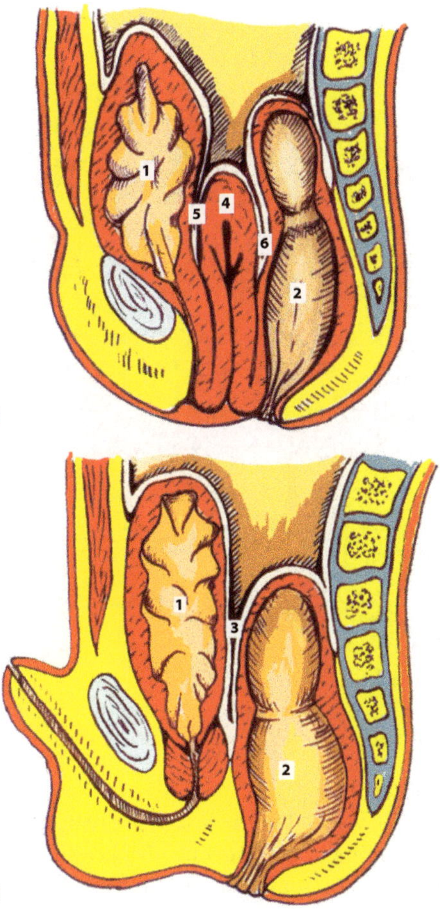

Figure 7 Topography of the bladder.
a – male, b – female:
1 – vesica urinaria; 2 – rectum; 3 – excavatio rectovesicalis; 4 – uterus; 5 – excavation vesicouterina; 6 – excavatio rectouterina

Syntopy of the bladder. The anterior surface of the bladder covered with a visceral fascia, is attached to the superior branches of the pubic bones and the pubic symphysis, separated from them by a layer of loose connective tissue of the retropubic space. The ampulla of the vas deferens, the seminal glands, the terminal sections of the ureters and fascia rectoprostatica (septum rectovesicale) adjoin the posterior surface of the bladder. Over a certain length, the vas deferens and ureters, crossing the vas deferens from the bottom and from the outside,

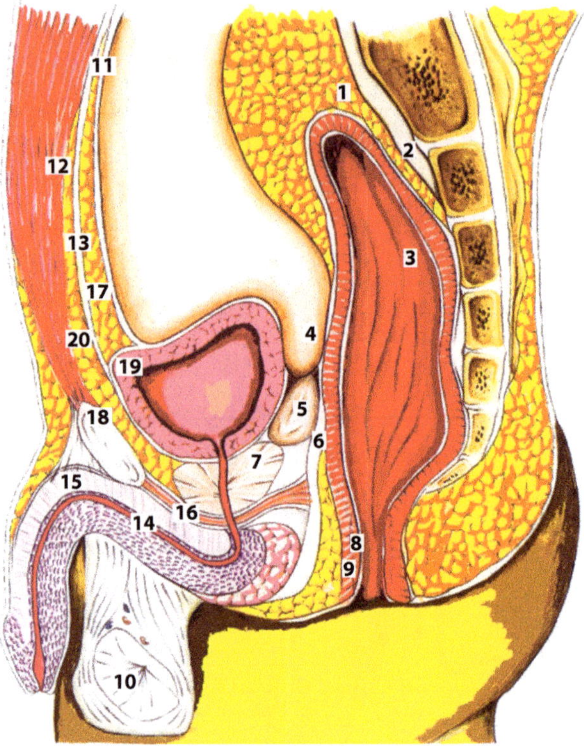

Figure 8 Sagittal section of the male pelvis.
1 - panniculus pararectalis; 2 – visceral pelvic fascia; 3 – rectum; 4 – excavatio rectovesicalis; 5 – vesiculae seminalis; 6 – aponeurosis peritoneoperinealis; 7 – prostata; 8 – m. sphincter ani internus; 9 – m. sphincter ani externus; 10 – testis; 11 – peritoneum; 12 – m. rectus abdominis et fascia transversalis; 13 – fascia prevesicalis; 14 – urethra; 15 – corpus cavernosum penis; 16 – diaphragma urogenitalis; 17 – panniculus preperitonealis; 18 – symphysis; 19 – vesica urinaria; 20 – panniculus prevesicalis

adjoin the lateral surfaces of the bladder. The loops of the intestines, the sigmoid, and sometimes the transverse colon or the cecum with the appendix adjoin the bladder at the top and from the sides. The fundus of the bladder adjoins the prostate.

The blood supply of the bladder. (Figures 8, 12 and 13) It carries out from the system of a. iliaca interna. One or two a. vesicalis superior mostly depart from unobliterated part of a. umbilicalis, a. vesicalis inferior – directly from the front of the trunk of a. iliaca interna or the obturator artery.

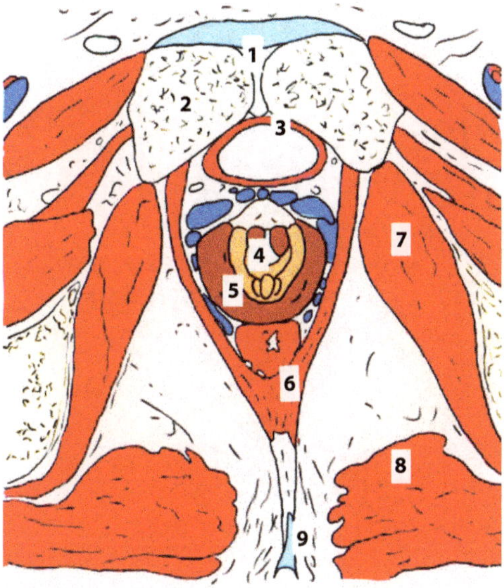

Figure 9 Cross section of the male pelvis.
1 – symphysis pubis; 2 – os pubis; 3 – vesica urinaria; 4 – urethra; 5 – prostate; 6 – anus; 7 – m. obturatorius internus; 8 – m. gluteu maximus; 9 – lig. anococcygeum

Veins of the bladder form a network in the visceral cellular space of the bladder. From there, the blood is sent to the venous plexus of the bladder and the prostate, located in the retropubic space. Then the blood flows into v. iliaca interna.

The lymphatic drainage of the bladder. It is carried out in nodi lymphoidei iliaci, located along the external iliac arteries and veins, and in nodi lymphoidei iliaci interni and sacrales.

The innervation of the bladder. The superior and inferior hypogastric nerve plexuses, the pelvic internal nerves and the genital nerve innervate the bladder. They form plexus vesicalis on the walls of the bladder and especially at the place of the entry of ureters into it.

The children features (Figures 7 and 11). In newborns and children, the topography of the pelvic organs is significantly different from adults. The bladder is mostly above the symphysis, its anterior wall is not covered by the peritoneum and is attached to the anterior abdominal wall. The urinary duct goes from the superior wall of the bladder to the navel. It quickly empties and obliterates, turning into a

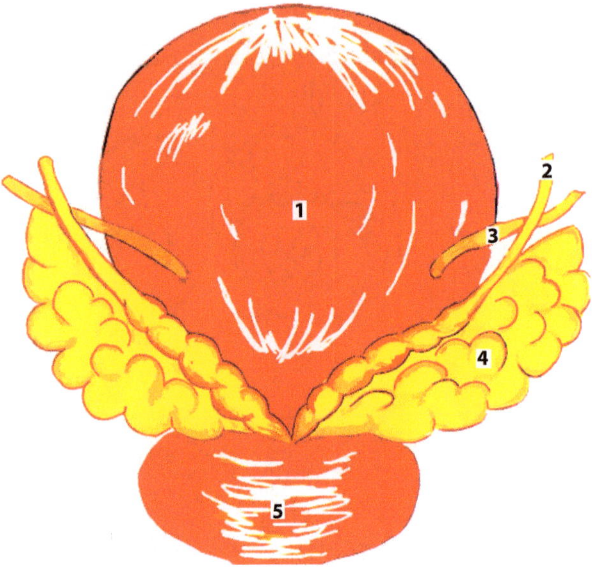

Figure 10 The bladder and the prostate.
1 – vesica urinaria; 2 – ductus deferens; 3 – ureter; 4 – vesicula seminalis; 5 – prostata

connective tissue cord. With age, the cavity of the small pelvis of the child increases, and the bladder descends and in the emptied state is located in the cavity of the small pelvis behind the symphysis.

Topography of the ureters. The ureter is a paired organ, it has three narrowing in its length: at the beginning of the ureter, at the place of transition of the abdominal part of the ureter into the pelvic part and at the point of influx into the bladder.

The pelvic ureter, which is about half its length, starts from the borderline of the pelvis. At the level of this line, the left ureter crosses the common iliac artery, and the right iliac – the external iliac artery.

Further, the ureter is located on the lateral wall of the pelvis in the lateral cellular space medially from the nerve trunks and the internal iliac vessels of the pelvis and laterally from the rectum.

Then the ureter crosses the obturator neurovascular bundle and the beginning of the umbilical artery and is directed medially to the fundus of the bladder.

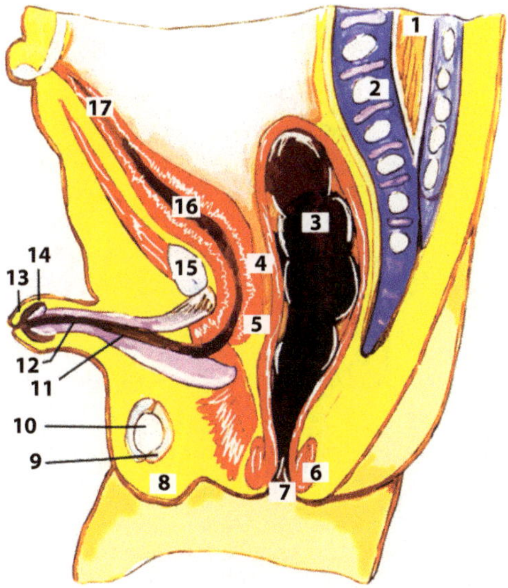

Figure 11 Median sagittal section of the pelvis of a newborn boy.
1 – cauda equine; 2 – os sacrum; 3 – rectum; 4 – vesicula seminalis; 6 – prostate; 7 – sphincter ani externus; 8 – anus; 9 – scrotum; 10 – epididymis; 11 – testis; 12 – urethra; 13 – corpus spongiosum penis; 14 – preputium; 15 – symphisis pubica; 16 – vesica urinaria; 17 – urachus

Here the ureter passes between the posterior wall of the bladder and the anterolateral wall of the ampulla of the rectum and crosses at right angles to the ductus deferens, located outside of it and anterior to the seminal glands.

The blood supply to the pelvic ureter is carried out from aa. rectales mediae and aa. vesicales inferiores.

Venous blood flows into vv. testiculares and vv. iliacae internae.

Pelvic parts of the ureters are **innervated** from the superior and inferiior hypogastric plexuses, while in the lower parts receives parasympathetic innervation from nn. splanchnici pelvini.

Outflow of lymph goesfrom the pelvic parts of the ureters to the iliac lymph nodes.

Topography of the prostate. The prostate consists of 30–50 glands forming the substantia glandularis, and a muscle substance (substantia muscularis) representing the stroma of the gland. The glands through the ductuli prostatici open into the prostatic part of the urethra. The prostate

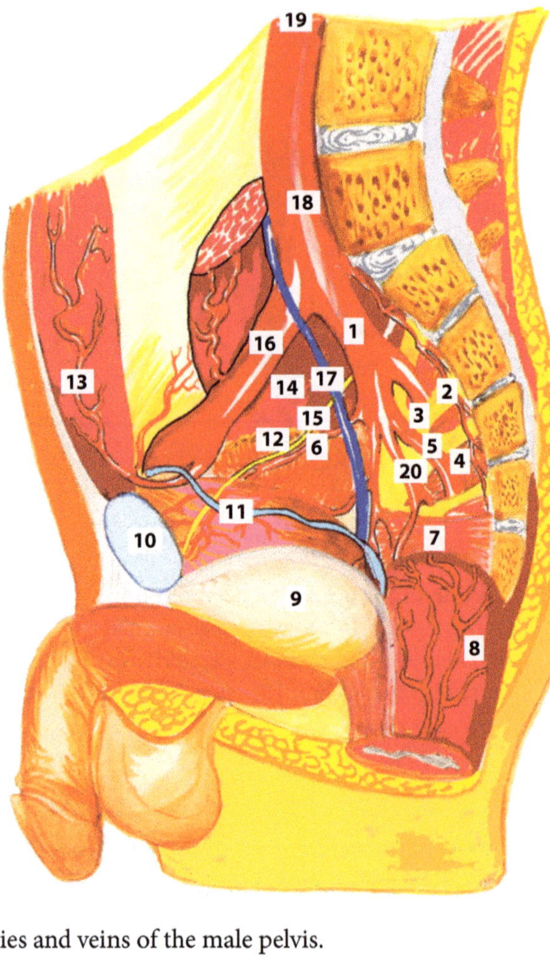

Figure 12 Arteries and veins of the male pelvis.
1 – a. iliaca interna; 2 – a. sacralis media; 3 – plexus sacralis; 4 – m. piriformis; 5 – a. glutea; 6 – a. umbilicalis; 7 – m. levator ani; 8 – rectum; 9 – vesica urinaria; 10 – os pubiis; 11 – ductus deferens; 12 – os ilium; 13 – a. epigastrica inferior; 14 – m. iliacus; 15 – n. obturatorius; 10 – a. iliaca externa; 17 – urethra; 18 – a. iliaca communis; 19 – aorta abdominialis; 20 – a. pudenda interna

is located in the subperitoneal floor of the small pelvis. It has a conical shape and is directed vertically downwards to the urogenital diaphragm. The base of the prostate is located beneath the fundus of the bladder. In the prostate, two lobes and an isthmus are distinguished. The prostate has a visceral fascial capsule (capsula prostatica or Pirogova-Retzia), from which mm. (ligg.) puboprostaticae go to the pubic bones.

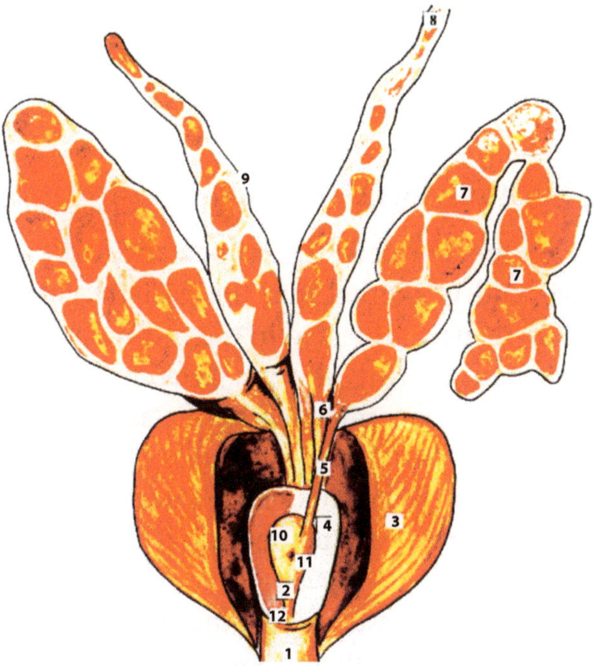

Figure 13 The prostate and the seminal vesicles.
1 – pars membranacea uretrae; 2 – tuberculum seminalis; 3 – prostate; 4 – pars prostatica urethrae; 5 – ductus ejaculatorius; 6 – ductus excretorius vesiculae seminalis; 7 – vesiculae seminales; 8 – ductus deferens; 9 – ampulla ductus deferentis; 10 – ductus deferens; 11 – utriculus prostaticus; 12 – crista urethralis

Syntopy of the prostate (Figure 13).

There are the fundus of the bladder, the seminal glands and the ampulla of the vas deferens above the prostate.

Beneath the prostate there is the urogenital diaphragm, in the front there is the posterior surface of the pubic symphysis, and behind there is the fascia rectoprostatica (Denonvilliers -Salishcheva) and the ampulla of the rectum. The prostate is easily pulpable through the rectum.

The blood supply of the prostate is carried out by branches of aa. vesicales inferiores and aa. rectales mediae (from a. iliaca interna).

The veins form the venous plexus (the plexus prostaticus) which merges with the plexus vesicalis; then the blood flows into v. iliaca interna.

Innervation is carried by the branches of the inferior hypogastric plexus.

The outflow of lymph from the prostate is carried out to the lymph nodes located along a. iliaca interna, a. iliaca externa and the pelvic surface of the sacrum.

The Topography of the Vas Deferens.

The pelvic part of the vas deferens is located in the subperitoneal floor of the pelvis and is divided into the parietal, intermediate and cystic parts. The cystic part is located in the retrovesical cellular space. Leaving the deep inguinal ring, the vas deferens round outside and down a. epigastrica, first accompanied by the same name artery, and then without it. Having rounded a. et v. iliacae externae, the vas deferens goes medially and backwards to the lateral space of the pelvis. Here, it crosses the obturator vascular bundle, the umbilical artery and the superior cystic arteries.

Located inwardly from these vessels, the vas deferens reaches the lateral wall of the bladder, and then passes between the ureter and the posterior surface of the bladder, forming an ampoule of the vas deferens (ampulla ductus deferentis). On the posterior wall of the bladder, the ampoule is located inwardly from the ureter and the seminal gland. The duct of the ampoule merges with the duct of the seminal gland (ductus excretorius) and forms the ejaculatory duct (ductus ejaculatorius), which penetrates the prostate and opens on the seminal tubercle into the prostate part of the urethra. The vas deferens are supplied with blood from a. ductus deferentis.

The topography of the testicles (Figure 13). The testicles are saccular protrusions outside the terminal sections of the ductus deferens. They are surrounded by the visceral fascia and located between the posterior wall of the bladder and the anterior wall of the ampulla recti.

Syntopy of the testicles. In front of the testicles there are the posterior wall of the bladder and the terminal section of the ureters. Medially they are attached to the vas deferens with ampoule. The lower parts of the testicles lie on the base of the prostate, and the upper medial sections are covered with the peritoneum, through which they adjoin the loops of the intestine.

Blood supply of the testicles is carried out from aa. vesicalis inferior et rectalis media. The veins fall into plexus vesicalis.

The testicles are **innervated** by the inferior hypogastric plexuses.

Lymph outflow from the testicles goes through the lymphatic vessels of the bladder to the lymph nodes located along the iliac arteries and the sacrum.

The children features. In boys, the prostate gland and seminal vesicles are also located relatively high compared to their position in adults.

The topography of the rectum. The rectum (rectum) is a continuation of the sigmoid colon and is located within the small pelvis at the anterior surface of the sacrum. In the rectum, the supraampullar part and the ampoule are distinguished. The supraampullar part and the upper half of the ampoule are located in the peritoneal floor of the small pelvis. The lower half of the ampulla of the rectum is located in the subperitoneal floor of the pelvis and instead of the peritoneum it is covered with the visceral fascia (Amussat capsule). The pelvic part of the rectum due to the curvature of the sacrum and coccyx forms a bend directed by the bulge back (flexura sacralis). With the transition to canalis analis, the ending part of the rectum deviates down and back, forming a second bend, anorectal flexure, facing the bulge forward (flexura anorectalis s. flexura perinealis). The rectum makes three bends in the frontal plane. These include the superior right lateral flexure (flexura superor dextra lateralis), the intermediate left lateral flexure (flexura intermedia sinistra lateralis) and the inferior right lateral flexure (flexura inferor dextra lateralis).

The layers of the rectum. Muscular layer consists of an external longitudinal (stratum longitudinale) and internal circular (stratum circulare) layers. At the level of the diaphragm of the pelvis over the fibers of the external sphincter (m. sphincter ani externus), muscle fibers of m. levator ani are interlaced in the muscles of the rectum, in particular, m. puboanalis and m. puborectalis. The mucosa of the upper part of the ampulla of the rectum forms 2–4 folds that do not disappear when the rectum is filled. The folds have a helical course. In the ampullar part, there is one fold on the right wall and two on the left.

Syntopy of the rectum. In the subperitoneal floor, anterior to the rectum there are the posterior wall of the bladder uncovered by peritoneum, the prostate, the ampulla of the vas deferens, the testicles (vesicles) and the terminal sections of the ureters. The rectum is separated from them by fascia rectoprostatica (Denonvillier-Salishcheva or septum rectovesicale). On the sides of the ampoule, there are the branches of the internal iliac vessels and the nerves of the plexus hypogastricus inferior. At the back, the rectum borders the sacrum.

Blood supply. A.rectalis superior (unpaired) is the terminal branch of the inferior mesenteric artery and a. rectalis media (paired, from a. iliaca interna).

Venous outflow. The veins form venous plexus in the wall of the rectum, (plexus venosus rectalis), in which the submucosal and subfascial parts are distinguished. From the upper parts of the rectum the blood flows through v. rectalis superior, which is the beginning of v. mesenterica inferior (the portal venous system). All veins widely anastomose with each other and with the veins of other organs of the pelvis. Thus, there is one of the portacaval anastomoses.

The inferior mesenteric, the superior and inferior hypogastric plexuses and the internal pelvic nerves **innervate** the rectum. The sacral spinal nerves include sensitive nerves that transmit a sense of filling the rectum.

The outflow of lymph from the supraampullar part of the rectum and partly from the upper part of the ampulla is carried out through the nodi pararectales along the superior rectal artery in the nodi rectales superiores and further in the nodi mesenterici inferiores. This explains the possibility of spreading cancer metastases along the rectum. From the peritoneal part of the rectum the lymph goes to the internal iliac and sacral lymph nodes.

The children features. The rectum in newborns is located high, it is expanded and its flexures are poorly expressed. It adjoin the ureters, the bladder and the vagina (in girls), the prostate gland and the testicles (in boys). With age, the topographo-anatomical relations of the rectum approach those in adults.

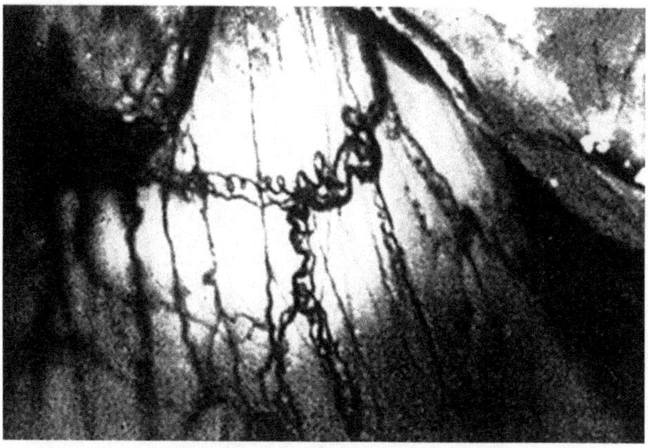

Figure 14 Transillumination angiogram of the broad ligament.

The Organs of the Female Pelvis

The topography of the bladder in women. The bladder in the female pelvis lies deeper in the pelvic cavity than in men. In the peritoneal floor, behind the bladder there are the corpus of the uterus and the intestinal loops, which enter the excavatio vesicouterine. In the subperitoneal floor, the bladder adjoin the pubic symphysis by the anterior surface and fixes to it by the pubovesical muscles/ligaments (mm. (ligg.) pubovesicalia). The posterior wall of the bladder lies anteriorly to the cervix and vagina. The bladder is firmly connected with the vagina, separating from it only with a small layer of fatty tissue; the fusion is more friable with the uterus. Laterally to it there is m. levator ani. At the fundus of the bladder in front of the anterior wall of the vagina, the ureters enter into the bladder. Lymphatic vessels of the bladder in women are associated with the lymphatic vessels of the uterus and the vagina at the base of the broad ligament of the uterus.

The topography of the uterus and its appendages (Figures 14 and 15). The uterus is located in the small pelvis between the bladder in front and the rectum from behind. It consists of two parts: the upper – the corpus (corpus) and the fundus (fundus) and the lower – the cervix uteri (cervix uteri). The supravaginal portion of cervix and vaginal one (portio supravaginalis et portio vaginalis) are distinguished in the cervix uteri. In portio vaginalis cervicis there is the opening of the

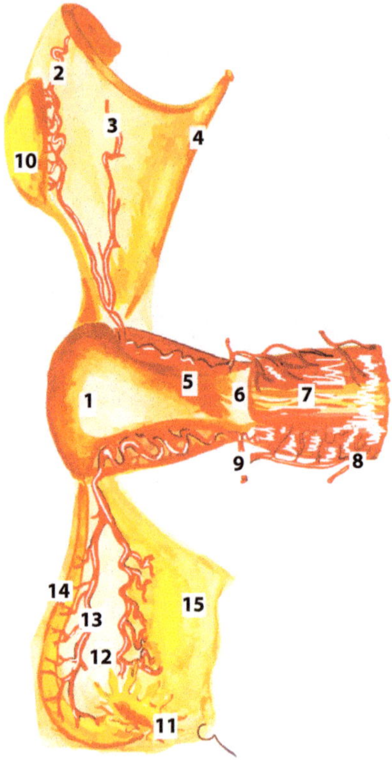

Figure 15 Blood supply of female reproductive organs.
1 – fundus uteri; 2 – a. ovarica; 3 – rr. ovarici a. uterinae; 4 – lig. teres uteri; 5 – corpus uteri; 6 – cervix uteri; 7 – vagina; 8 – rr. vaginales a. uterinae; 9 – a. uterine; 10 – ovarium; 11 – infundibulum tubae uterinae cum fimbriac; 12 – mesenterium tubae uterinae; 13 – ramus tubaria a. uterinae; 14 – tuba uterine; 15 – mesenterium ovarii

uterus, (ostium uteri) limited in front by labium anterius and behind by labium posterius. This opening connects through the canalis cervicis uteri vagina with the uterine cavity (cavum uteri). The lateral edges of the uterus are called margo uteri dexter et sinister. Most of the uterus is located in the peritoneum floor of the small pelvis.

The suspensory apparatus of the uterus is formed by round and broad uterine ligaments (ligg. teres uteri et ligg. lata uteri). The broad ligaments of the uterus are duplications of the peritoneum. They depart from the uterus almost in the frontal plane and reach the peritoneum of the lateral walls of the pelvis. At this place, the peritoneal sheets of the broad ligament form the suspensory ligament (lig. suspensorium ovarii), containing the ovarian vessels (a. et v. ovarica). Downward and backward from the angle of the uterus in the thickness of the broad ligament its own ligament of the ovary (lig. ovarii proprium) is departed. Downward and anteriorly to the angle of the uterus there is the round ligament of the uterus (lig. teres uteri). The uterus has the visceral fascia. Musculo-fibrous bundles of the main ligament of the uterus (lig. cardinale) go to the direction of the lateral wall of the pelvis from the fascia on the sides at the base of the broad ligament of the uterus. Ligaments that attached to the visceral fascia: the cardinal (ligg. cardinalia), the recto-uterine (ligg. rectouterine), pubocervical (ligg. pubocervicale) for fixing the vagina, as well as the pubovaginal muscle (m. pubovaginalis), the urethrovaginal sphincter (m. sphincter urethrovaginalis) and the perineum membrane (membrana perinei).

Blood supply (Figures 16 and 17) is carried out by two uterine arteries (aa. uterinae), ovarian arteries from the abdominal aorta (aa. ovaricae) and the round ligament arteries (aa. lig. teretis uteri). A. uterina is a branch of the internal iliac artery.

Venous outflow from the uterus is firstly made in the uterine venous plexus (plexus venosus uterinus). It widely anastomoses with all the veins of the pelvis, but primarily with the venous plexus of the vagina (plexus venosus vaginalis). From the plexus, the blood flows through the uterine veins into the internal iliac veins. The venous blood from the fundus of the uterus, the ovaries and the tubes goes to the inferior vena cava via vv. ovaricae.

The innervation of the uterus is carried out by an extensive utero-vaginal neural plexus (plexus uterovaginalis) – the middle portion of the inferior hypogastric plexus (plexus hypogastricus inferior).

The outflow of lymph. From the uterus through the parauterine lymph nodes and the paravaginal lymph nodes (nodi parauterini et paravaginales), the lymph flows into the iliac lymph nodes and further into the common iliac nodes. Along ligg. cardinalia lymph is carried to the lymph nodes,through the cervix lymphatic vessels and then to the external and common iliac nodes. From the fundus of the uterus, the lymph partly flows to the inguinal lymph nodes through the lymphatic vessels of the round ligament of the uterus.

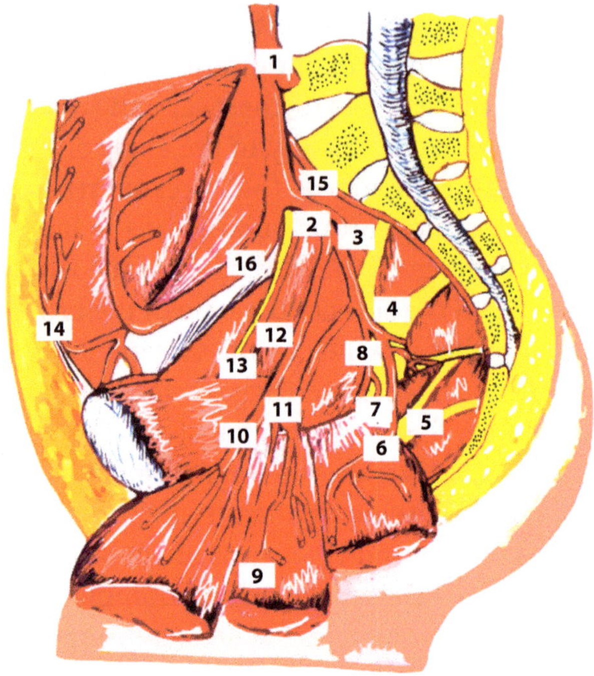

Figure 16 Arteries and veins of the female pelvis.
1 – a. iliaca communis; 2 – a. iliaca interna; 3 – a. glutea superior; 4 – plexus sacralis; 5 – plexus pudendus; 6 – a. glutea inferior; 7 – a. rectalis media; 8 – a. pudenda interna; 9 – a. uterine; 10 – a. cystica; 11 – margo fascia pelvina; 12 – m. piriformis; 13 – a. et n. obturatoria; 14 – a. epigastrica inferior; 15 – a. sacralis externa; 16 – a. iliaca communis

Fallopian tubes (Figures 14, 15 and 19). They also known as uterine tubes or salpinges (tuba uterine). Fallopian tubes are paired organ that connects the uterine cavity with the peritoneal cavity. It is located along the superior edge of the broad ligament of the uterus and has a mesentery (mesosalpinx), which is part of a broad ligament. The diameter of the Fallopian tube is not the same and varies from 5 to 10 mm. The uterine portion (uterine pars) with the uterine opening (the ostium uterinum), the isthmus (isthmus), the ampulla (ampulla), and the infundibulum (infundibulum) are distinguished in the Fallopian tube. The infundibulum has fimbriae (fimbriae), fringing the abdominal opening of the tube (ostium abdominale tubae uterinae). One of the fringes is located at the tube end of the ovary and is called fimbria ovarica.

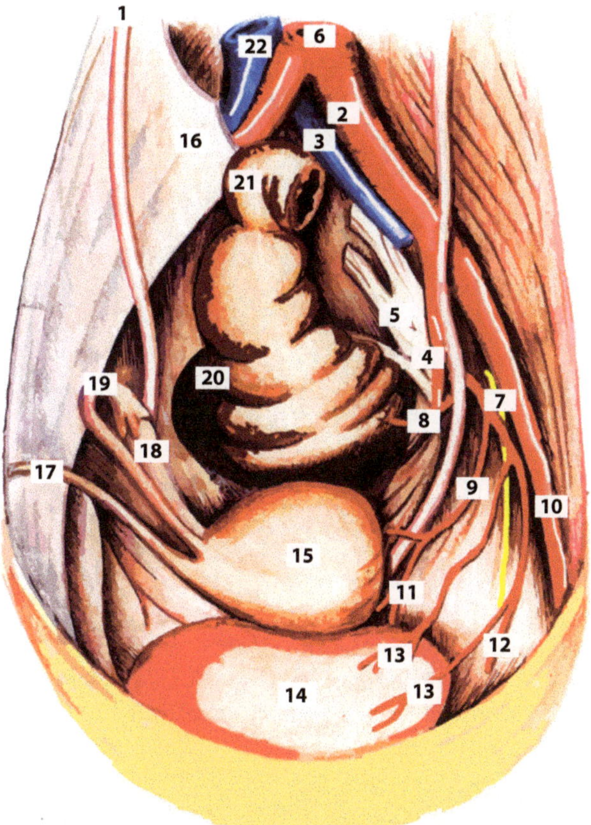

Figure 17 The topography of the female pelvis. The uterus and vagina at the base of the broad ligament of the uterus.
1 – uterus; 2 – a. iliaca communis; 3 – v. iliaca communis; 4 – a. iliaca interna; 5 – plexus sacralis; 6 – aorta; 7 – n. obturatorius; 8 – a. rectalis media; 9 – a. uterine; 10 – a. iliaca externa; 11 – a. vaginalis; 12 – a. umbilicalis; 13 – a. cystica; 14 – vesica urinaria; 15 – uterus; 16 – peritoneum; 17 – lig. teres uteri; 18 – ovarium; 19 – tuba uterinae; 20 – lig. sacrouterina; 21 – rectum; 22 – v. cava inferior

Blood supply (Figure 16) of the Fallopian tubes is carried out from the ovarian and uterine arteries.

Ovaries. Ovaries are paired female glands measured 1.5 × 1.5 × 1.0 cm. They are covered with a germinal epithelium. The transition of the epithelium to the endothelium of the peritoneum is marked with a white line. At this point, the mesentery of the ovary (the mesovarium), that goes from the posterior leaf of the broad ligament of the uterus,

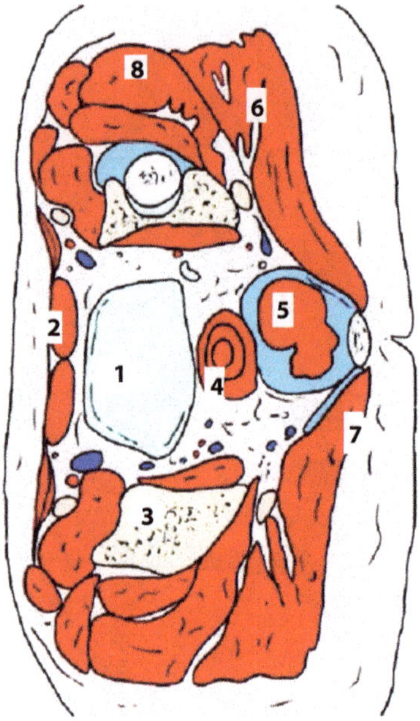

Figure 18 Sagittal section of the female pelvic organs.
1 – vesica urinaria; 2 – m. rectus abdominis; 3 – acetabulum; 4 – uterus; 5 – rectum; 6 – m. gluteus maximus; 7 – lig. sacrospinale; 8 – m. gluteus maximus

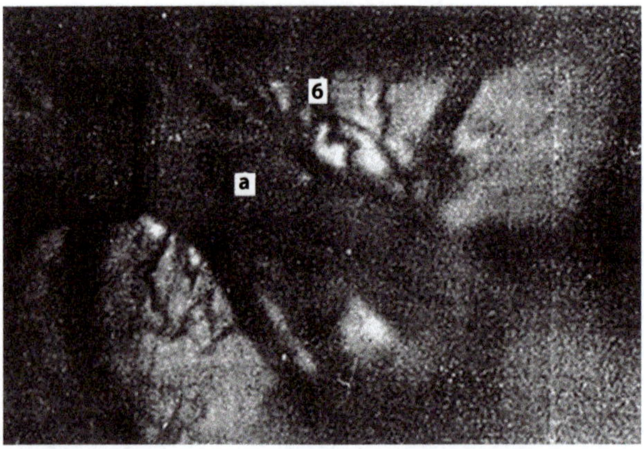

Figure 19 The round ligament of the uterus in transmitted light.
1 – ligamentum; 2 – vasa subperitonealis

ends. Ovary has two extremities – tubal and uterine, two surfaces – medial and lateral, and two borders – free and mesenteric. The ovary is attached to the posterior surface of the broad ligament of the uterus at the lateral wall of the pelvis. Under the peritoneum a. ovarica goes from the retroperitoneal space to the ovary. The uterine extremity of the ovary is connected to the uterus by the connective tissue ligament of the ovary (lig. ovarii proprium).

The blood supply of the ovary is carried out from a. ovarica, extending from the abdominal part of the aorta at level of the I lumbar vertebra, as well as the ovarian branch of the uterine artery.

The outflow of venous blood from the ovary occurs through v. ovarica dextra directly into the inferior vena cava, through v. ovarica sinistra – firstly to the left renal vein and through it to the inferior vena cava.

In the innervation of the ovary, the branches of the inferior hypogastric plexus participate.

Lymph outflow from the ovary is carried out by the lymphatic vessels accompanying the ovarian artery, into the lymph nodes located around the aorta, and then into the iliac lymph nodes.

Children features (Figure 17). In newborn girls, the uterus is located above the plane of entry into the small pelvis. It has not completed its development, its corpus is 1/3, and the cervix is 2/3 of the entire length. The ovaries are located near the borderline of the pelvis.

With age, the uterus and ovaries descend and by the age of 12–14, they occupy a position corresponding to their position in women. By this age, the size of the uterus increases, and the length of its corpus and the cervix becomes the same.

The topography of the rectum in women. The structure, division into portions, blood supply and innervation of the rectum in women do not differ from those in men. Only the syntopy and lymphatic drainage of the rectum are different. In the peritoneal floor, the corpus, the cervix and the posterior vaginal vault are located anteriorly to the rectum in women. Between the rectum and the posterior wall of the uterine body, there are the loops of the small intestine descending from the inferior floor of the abdominal cavity. They enter the recto-uterine pouch (excavatio rectouterine). In the subperitoneal floor, the rectum is in the front of the vagina, but they are separated by fascia. This fascia is rather thin and friable; it is pierced by lymphatic vessels, so it does not become an obstacle either for the spread of metastases of tumors of both organs or for the development of rectal-vaginal fistulas.

Defects of the Genitourinary System in Children

Cysts of the urinary duct (uracus), formed with its incomplete obliteration, sometimes have a fistulous course, opening in the navel - vesico-umbilical fistula.

There is also a congenital fistula wich is called vesico-intestinal fistula, but is extremely rare. It usually occurs between the rectum and vesical triangle region, sometimes combined with atresia of the anus.

In addition, in the urinary duct of girls, an ectopic insertion of the endometrial elements can occur. In these cases, at puberty in the menstrual period in the wisp remaining from the urinary duct, cysts can form filled with blood. In fistula the discharge of blood from the navel is possible.

Bladder exstrophy. This malformation is characterized by the absence of the anterior septum of the bladder and part of the anterior abdominal septum. The bladder is opened anteriorly, respectively mucosa defect of the bladder septum adherent to the edges of the skin defect. Pinhole ureters are clearly visible on the posterior septum of the bladder mucosa. Urine is continuously fed out from them.

Hypospadias - malformation characterized by the absence of the bottom septum of the urethra.

Epispadias - underdevelopment of the upper septum of the urethra.

Malformations of the rectum. Atresia of the anus (atresia ani). There is no the anus opening and rectum ends blindly close to the skin of the perineum. The rectum is usually stretched by accumulated in it meconium.

With rectal atresia, atresia recti, the anus is represented by a expressed depression, but the rectum is short and blindly terminates above the pelvic floor. The blind end of the rectum is separated from the perineum by a considerable layer of tissues.

With atresia of the anus and rectum, atresia ani et recti, which occurs more often than other developmental malformations, the anus remains closed and the rectum blindly ends at a different distance from the bottom of the small pelvis.

Perineum Topography

Boundaries, regions (Figures 20 and 21). The crotch region (regio perinealis), which forms the bottom septum of the pelvic cavity, has a diamond shape and is bounded by the pubic symphysis in the front, the front and laterally - the lower branch of the pubic bone and the sciatic branch, laterally – buttock, lateral and back – sacro-tubercular ligaments, posterior – coccyx. The line that connects the ischial tuberosity (linea biischiadica), the perineum is divided into urogenital and anal region in both men and women almost equally.

In the center of the anus region is the opening of the rectum (anus).

Layers of Perineal Region and Their Characteristics:

1. The skin (derma) of the anus is colored, thin and adherent to the subcutaneous part of the external anal sphincter. It forms folds, and then passes into the rectum. In men between the radix of the scrotum and the anus there is the crotch seam (raphe perinei);
2. Subcutaneous adipose tissue and superficial fascia of the anus (panniculus adiposus, fascia perinei superficialis) are expressed better than the genitourinary region. Cutaneous branches of the lower gluteal and lower rectal arteries and saphenous venous net, particularly thickened around the anus, are located in the fatty tissue. The branches of nn. rectales inferiores n. pudendi in the medial region and n. perineales of n. cutaneus femoris posterior in the lateral region innervate the skin;
3. corpus adiposum fossae ishiorectalis;
4. fascia diaphragmatica pelvis inferior;
5. m. levator ani;
6. fascia diaphragmatica pelvis superior;

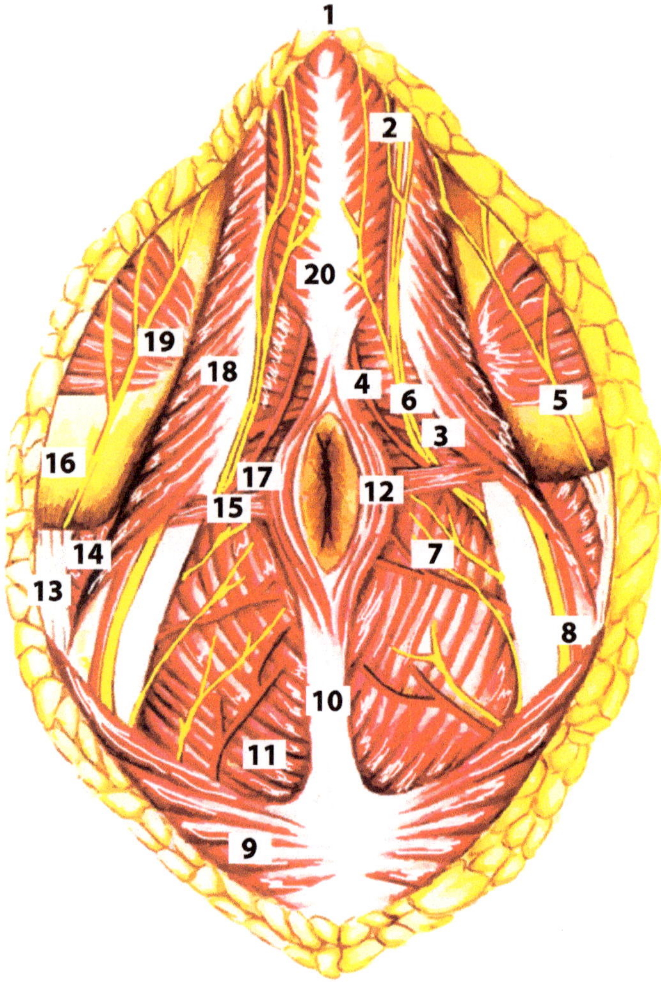

Figure 20 Topography of the male perineum.
1 – scrotum; 2 – aa. et nn. scrotales posteriores; 3 – a. perinealis; 4 – a. penis; 5 – r. perineus n. cutanei femori posterior; 6 – n. perineus; 7 – aa. et nn. rectales inferiores; 8 – vasa et nn. pudendales internus; 9 – m. gluteus maximus; 10 – lig. rectococcygeus; 11 – m. levator ani; 12 – m. sphincter ani externus; 13 – lig. sacrotuberositas; 14 – m. obturatorius internus; 15 – m. perinei transversus superfiecialis; 16 – tuber ischiadicus; 17 – m. perinei transversus profundus; 18 – m. ischiocavernosus; 19 – m. abductor; 20 – m. bulbocavernosus

7. cavum pelvis subperitoneale;
8. peritoneum parietale;
9. cavum pelvis peritoneale.

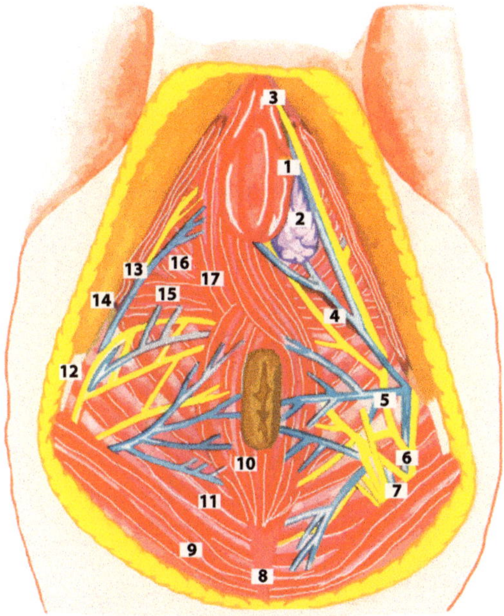

Figure 21 Topography of the female perineum.
1 – nn. et aa. labiorum pudendae; 2 – bulbus vestibuli; 3 – nn. el. aa. clitoridis; 4 – a. bulbi veslibuli; 5 – n. et. a. perinei; 6 – aa. et n. pudendes internes; 7 – aa. et n. rectales inferiores; 8 – lig. coccygioanalis; 9 – m. gluteus maximus; 10 – m. sphincter ani externus; 11 – m. levator ani; 12 – lig. sacrotuberalis; 13 – m. obturatorius internus; 14 – tuber ischiadicus; 15 – m. transversus perinei superficialis; 16 – m. transversus perinei profundus; 17 – m. bulbocavernosus

Bunches of vessels and nerves, cellular spaces are the perineal nerves (nn. perinealis), perineal branch of the posterior femoral cutaneous nerve (rr. perinealis n. cutaneus femori posterior), cutaneous branches of the inferior gluteal (a. et v. glutea inferior), and rectal (a. et v. rectalis inferior).Cellular spaces: parietal and visceral cellular spaces.

Pudendal Region in Men

Penis (penis) (Figures 7, 9 and 12) is located in front of the symphysis and tightly fastened to it by a sling ligament – lig. fundiforme penis. The suspending ligament (lig. Suspensorium) of the penis takes part in the fixation of it. The posterior end, the radix (radix penis) is attached by the bulb (bulbus) to the urogenital diaphragm and to the inside of the pubic and ischial bones – by the fusion of the gallbladder with the periosteum.

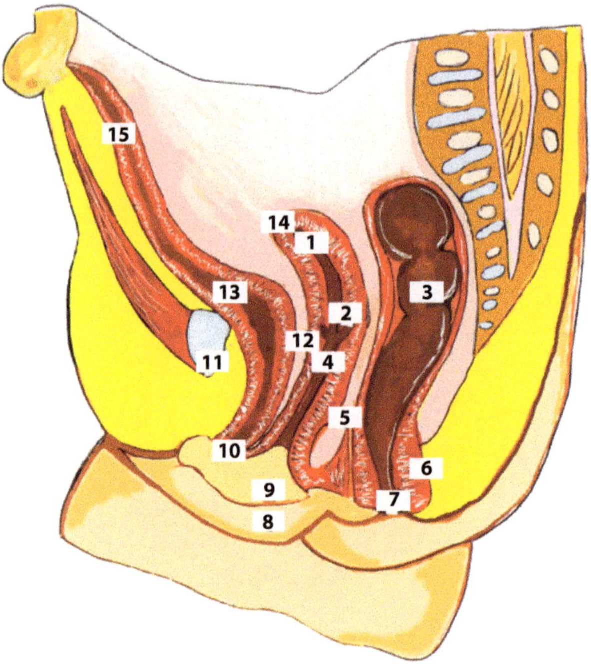

Figure 22 Median sagittal section of the newborn girl pelvis. A right side view.
1 – cavitas uteri; 2 – fornix vaginae (pars posterior); 3 – rectum; 4 – vagina; 5 – exavatio rectouterina; 6 – sphincter ani externus; 7 – anus; 8 – labium majus pudendi; 9 – labium minus pudendi; 10 – ostium urethrae externum; 11 – symphysis pubica; 12 – excavacio vesicouterina; 13 – vesica urinaira; 14 – uterus; 15 – urachus

The penis has a reinforced posterior end – the radix (radix penis), body (corpus penis), and head (glans penis); surfaces – the anterior (dorsum penis) and posterior (facies urethralis).

The structure of the penis consists of two erectile tissues (corpora cavernosa penis) and the spongy body (corpus spongiosum penis), which is thickened at both ends and form the head of the penis in the front and the bulb behind.

The skin that covers the body of the penis is thin and movable. The subcutaneous tissue of the penis (tela subcutanea penis) is friable, which provides an easy shift of the skin. In the subcutaneous tissue in the groove between the corpora cavernosa there is v. dorsalis penis, which heads to the venous plexus of the bladder. On either side of the vein there are aa. dorsales penis and more laterally – nn. dorsales penis.

Fascia of the penis is thick and surrounds all three cavernous bodies. On the back side of the penis beneath it there are paired corpora cavernosa. They are surrounded by a tunica (tunica albuginea corporum cavernosorum) and split septum of the penis. Each of the cavernous bodies begins from the branches of the pubic and sciatic bones forming a pedicle (crus penis), which is covered with sciatic-cavernous muscle (m. ishiocavernosus). In the thickness of the cavernous bodies of the penis there are deep arteries (aa. profundae penis), arising from aa. pudendae internae and with the help of perforating branches anastomosing with the back artery. Together with these arteries, the vegetative nerves (from the plexus hypogastricus) approach the penis.

There is an unpaired spongy body located beneath between the cavernous bodies of the penis. The posterior part of the spongy body (bulbus penis) is covered by m. bulbospongiosus and fixed to the membrana perinei (urogenital diaphragm). Crura and bulbus penis are the radix of the penis (radix penis). Cavernous bodies and the spongy body, covered with fascia penis, connect with the symphysis and form a corpus penis.

Blood supply of the penis is performed due to the internal pudendal artery (a. pudenda interna), which gives a way to a. dorsalis penis, that goes to the skin, cavernous bodies, the head and partly to the spongy body of the penis, and to a. profunda penis, that goes to the cavernous bodies.

Superficial veins of the penis go into the femoral vein and in the depth – into the inner pudendal vein. V. dorsalis penis carries blood to the venous plexus of the bladder.

Innervation of the penis is carried out by nn. dorsales penis (branch nn. pudendi). They are associated with autonomic nerve plexus pelvinus.

Lymphatic drainage from the penis is carried to the inner group of inguinal nodes and partly to external iliac lymphatic nodes.

Topography of the male urethra. Male urethra begins at the bottom of the bladder internal opening (ostium urethrae internum) and ends at the head of the penis outside the opening (ostium urethrae externum). Its length is 20–22 cm.

The male urethra has several parts:

1. Part of the internal septum (preprostatic part), pars intramuralis (pars preprostatica);
2. The prostatic part of the male urethra, pars prostatica;
3. The intermediate part of the male urethra (membranous part), pars intermedia (pars membranacea);
4. Spongy part of the male urethra (pars spongiosa).

The blood supply to the male urethra is provided by a. pudenda interna. The outflow of venous blood goes through the veins of the penis into the venous plexus of the bladder.

The innervation of the male urethra is from the genital plexus.

Lymphatic drainage from the male urethra is performed in the inguinal lymph nodes.

The topography of the scrotum. The scrotum is a cutaneous bag which is divided by a partition, in each half of which there are the testicle, its appendage and scrotal division of the spermatic cord surrounded by shells. The skin is collected in folds and covered with thin hairs. On the middle line it has raphe scroti.

Blood supply of the scrotum is performed by rr. scrotales anteriores (from aa. pudendae externae), posterior – rr. scrotales posteriores (from a. pudenda interna), as well as the artery muscles, suspensory testisle, a. cremasterica (from a. epigastrica inferior).

Venous blood flows away from the scrotum of the same name veins.

The scrotum is **innervated** anteriorly by nn. scrotales anteriores (from n. ilioinguinalis), posteriorly – nn. scrotales posteriores (from nn. perinealea, branches n. pudendus).

Lymph outflow from the scrotum is performed in the superficial inguinal lymph nodes.

Layered structure of the scrotum (Figure 23):

1. The skin (derma).
2. Fleshy sheath (tunica dartos).
3. External seminal fascia (fascia spermatica externa).

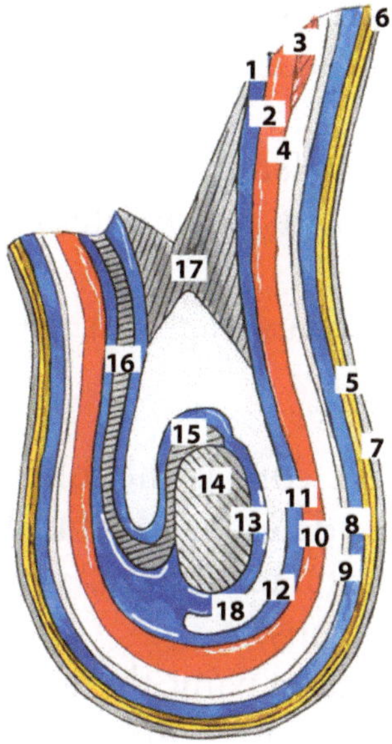

Figure 23 Shell of testis and layers of the scrotum.
1 – peritoneum; 2 – fascia transversa; 3 – mm. obliquus internus et transversus andominis; 4 – aponeurosis m. obliquus externus abdominis; 5 – panniculus subcutaneus; 6 – derma pariei abdominis anterior; 7 – derma scroti; 8 – tunica dartos; 9 – fascia spermatica externa; 10 – m. cremaster; 11 – fascia spermatica interna; 12 – lam. parietalis tun. vaginalis testis; 13 – tun. albuginea testis; 14 – testis; 15 – epididymis; 16 – ductus deferens; 17 – ligg. vaginalis; 18 – lam. visceralis tun. vaginalis testis

4. Fascia of the muscle lifting testicle (fascia cremasterica) is the fascia descending into the scrotum of the external fascia of the external oblique muscle of the abdomen (fascia propria m. obliqui abdominis externi).
5. muscle lifting testicle (m. cremaster).
6. internal spermatic fascia (fascia spermatica interna).
7. tunica vaginalis testis (tunica vaginalis testis).
8. perididymis (tunica albuginea testis).

The spermatic cord (funiculus spermaticus) stretches from the upper end of the testicle to the deep inguinal ring (annulus inguinalis profundus) and contains the following formation:

- the cord (pars funicularis) and groin (pars inguinalis) parts of the vas deferens (ductus deferens);

- testicular artery (a. testicularis);
- testicular vein (v. testicularis);
- lymphatic vessels;
- the artery of the vas deferens (a. ductus deferentis);
- cremasteric artery (a. cremasterica) and vein (v. cremasterica);
- vestige of the processus vaginalis (vestigium processus vaginalis).

Location of the spermatic cord is the following: in the back of its department there is the vas deferens (ductus deferens), in front of its there is the testicular artery (a. testicularis) and behind – the artery of the vas deferens (a. deferentialis).

The same name veins accompany the arterial trunks. Lymphatic vessels in large numbers are on the front group of veins. These formations cover the internal spermatic fascia (fascia spermatica interna), fascia muscle lifting testicle (fascia cremasterica) and external seminal fascia (fascia spermatica externa).

Pudendal Region in Women

The Topography of the External Female Genitalia

Large labia (labia majora pudendi) is formed of two cutaneous rollers restricts external pudendal cleft (rima pudendi). In the front of the pubis (mons pubis) labia lips are connected to the anterior commissure (comissura labiorum anterior). Posteriorly the labia is connected to the posterior commissure (comissura labiorum posterior) near the anal area.

Labia minora (labia minora pudendi) is located medially large labia and represents cutaneous rollers, which are connected by the frenulum (frenulum labiorum pudendi). Fossa vestibuli vaginae is located in front of the frenulum there. Labia minora are split into two parts in the front. One is located on top of the clitoris and forms his foreskin (preputium clitoridis). The other goes under the clitoris and forms its frenulum (frenulum clitoridis). Anteriorly the labia minora there is the vestibule (vestibulum vaginae). It is restricted on the sides by the internal surfaces of the labia minora, above – by the clitoris, at the bottom – by the frenulum of the labia minora. In front of the vestibule there is external urethral opening (ostium urethrae externum). At the back of it there is ostium vaginae. The boundary between the vestibule and the vagina is the hymen or its remains.

Clitoris (clitoris). The clitoris is located in the upper corner of the pudendal cleft just under the pubis, and his head appears here as a small papilla. It consists of two cavernous bodies (corpora cavernosa dextrum et sinistrum), their pedicles (crura clitoridis) is about 1 cm thick spliced with the inferior branches of the pubic bone and covered by ischial-cavernosus muscles of the perineum. A little before reaching the lower edge of the pubic symphysis, the cavernous bodies are joined to form the body of the clitoris (corpus clitoridis). Under the pubic arch it bends down and ends with a rounded head (glans clitoridis). The convex part of the clitoris is fixed to the pubis suspensory ligament of the clitoris. V. dorsalis clitoridis passes between the pubic arch and the place of the inflection of the clitoris. The bulbs of the vestibule (bulbus vestibuli) lie under the skin and the fat pad of the labia majora and bulbous-spongy muscles.

The bulbs of the vestibule (bulbus vestibuli) lie under the skin and the fat pad of the labia majora and bulbous-spongy muscles. The bulb of the vestibule is a venous plexus. It is intertwined with the fibers of the connective tissue and smooth muscle fibers and has form of the cavernous tissue. The superior narrowed part of the bulb covers the sides of the urethra and is connected by a thin arc comissure (comissura bulborum). It is located between the clitoris and urethra.

Large gland of the vestibule (glandula vestibularis major) is located at the base of the labia majora and at the back of adhesions in the thicknesses of urethrovaginal sphincter (m. sphincter urethrovaginalis) at a depth of 1.0–1.5 cm. It has the form of reddish-yellow body in the size of a large pea. Its excretory duct is directed forward and medially and opens on the internal surface of the labia minora.

The vagina is a tube. It consists of muscular and mucous membranes. It is located in the anterior part of the small pelvis between the bladder and the rectum. The superior extremity of the vagina surrounds the cervix. External opening of the vaginal part of the cervix is toward to the posterior septum of the vagina. Between the cervix and the superior extremity of the vagina it forms an annular groove (fornix vaginae). It has four parts: anterior, posterior and two side.

Syntopy of the vagina. The front septum of the vagina adjacents to the bottom of the bladder and urethra and tightly connects with them. As a result, the front septum of the vagina forms an elevation (carina urethralis vaginae). The posterior septum of the vagina adjacents to the rectum in the area from the peritoneum to the center of the perineum and is separated from it by fascia rectovaginalis or Denonvillier-Salishchev fascia (septum rectovaginale), which in women is weaker than the similar fascia rectoprostatica (septum rectovesicale) in men. From the sides of the vagina there is fatty tissue of base broad uterine ligament. On each side in the tissue next to the vagina thre are the ureters, uterine arteries, uterovaginal venous plexus, nodes and branches of the uterovaginal plexus, lymphatic vessels. Lower of the pelvic floor, on either side of the vagina internal borders of m. levator ani are located. Further below the vagina enters into the perineal floor and passes through uretrovaginal sphincter (m. Sphincter urethrovaginalis). Topographically it is an analogue of the deep transverse perineal muscles in men. It differs from it in that it consists of smooth muscle fibers. Fascial compartment of the vagina is closely coalesced to the muscle and membrane. The perineal floor of the vagina there is the opening (ostium vaginae) between labia minora in the vestibule (vestibulum vaginae).

The blood supply of the vagina is carried out by a. vaginalis, vaginal branches of a. uterina (the upper part of the vagina), and vesicalis inferior (to the sheath of the middle section) and a. pudenda interna (the lower part of the vagina).

Vaginal veins form a venous plexus (plexus venosus vaginalis). It goes into uterovaginal venous plexus. The outflow of venous blood and innervation are the same as in the uterus. The lower part of the vagina is innervated by branches of the n. pudendus.

The lymphatic vessels of the lower part of the vagina terminate in inguinal nodes, and from the upper two-thirds of diverting vessels goes along with the lymphatic vessels of the uterus to the iliac nodes located on the side septum of the pelvis.

The topography of the female urethra The urethra starts from the internal opening of the bladder neck. Its muscle bundles form the internal sphincter. This is intramural part of the female urethra. It corresponds to the level of the lower edge of the symphysis and is 1.5–2.0 cm posteriorly to it. Then the female urethra is directed downward and forward and passes through uretrovaginalny sphincter. At this level, the female urethra is surrounded by the external urethral sphincter (m. sphincter urethrae externus). Further, the female urethra pierces the perineal membrane (urogenital diaphragm) and opens into the vestibule between the clitoris and the vaginal opening the outer space (ostium urethrae externum).

The length of the female urethra is on average 3 cm. The posterior surface of the female urethra is connected with visceral fascia of the anterior vaginal septum. The vaginal septum forms a small elevation, urethral vaginal keel (carina urethralis vaginae).

Features of the shape and position of the pelvic organs in children. Age characteristics of the rectum. In newborns the rectum has a thin septum, the bends and folds of which are poorly expressed, in filling the meconium, it takes almost the entire volume of the cavity adjacent to the pelvic and ureter, prostate and seminal vesicles (or vagina). The emptyfied rectum moves from these organs.

Age characteristics of the bladder. Newborns have a spindle shape. A large part of it is above the symphysis. Urachus extends from the apex of the bladder to the navel. Spindle-shaped bladder saves up to 1.5 years to 10 years, it has an egg shape, and to 15 years – a form of the adult bladder.

Age characteristics of the uterus (Figures 24–27). A newborn has a length of the uterus of about 3 cm, and the length of the cervix twice the length of the body. Soon after birth, there is regression of the uterus

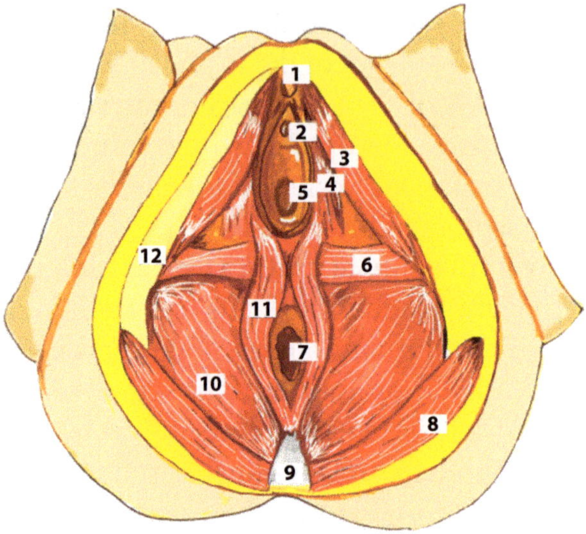

Figure 24 Topography of muscles and fascia perineum in a newborn girl.
1 – clitoris; 2 – ostium urethrae externum; 3 – mischiocavemosus; 4 – m. sphincter urethrovaginalis; 5 – ostium vaginae; 6 – m.transversus perinei superficial; 7 – anus; 8 – m. gluteus maximus; 9 – cortilago coccyges; 10 – m. levator ani; 11 – m.sphincter ani extemus; 12 – tuber ischiadicum

(length is reduced to 2 years to 2.5 cm). Further there is a slow growth due to the increase in the length of the uterine body.

By age 16, the uterus attains to a length of 6.6 cm. The uterus in infants and young children has an elongated shape, by 12–14 years it becomes pear-shaped like in women.

Surgical Anatomy of Congenital Pelvic and Perineum

Bladder Exstrophy is a severe malformation, in which the child does not have the front septum of the bladder and the corresponding anterior abdominal septum (abdominal septum). It occurs in boys 2 times more often girls.

The nature of this defect is still not clear. Exstrophy of the bladder is always accompanied by total epispadias and splitting of the pubic bone. There are 3 degrees of exstrophy, depending on the size of the defect of the anterior abdominal septum, the distance at which the costs of the pubic bone, and concomitant developmental defects.

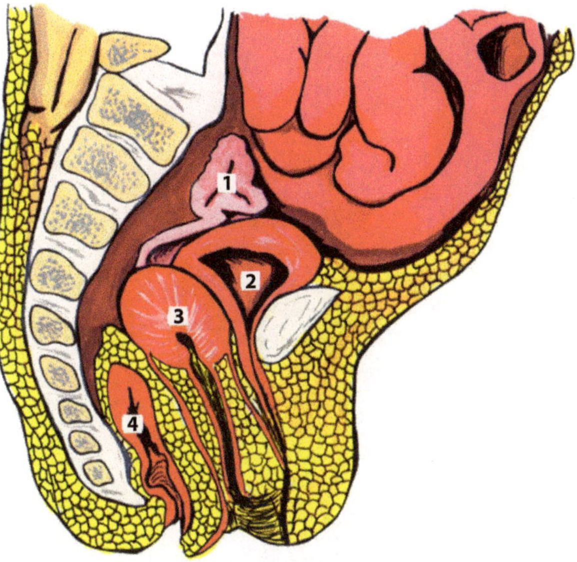

Figure 25 Sagittal section of pelvic girl 5 months old.
1 – ovarium sinister; 2 – vesica urinaria; 3 – uterus; 4 – rectum

The diagnosis is established at birth. Treatment of the bladder exstrophy is only surgery. This method depends on the severity. So at 1 degree exstrophy the defect of the anterior abdominal septum is surgically closed and the bladder is sutured, and at 2 and 3 degrees ureters are transplanted into the intestines, the bladder itself is removed and abdominal septum defect is sutured.

Diverticulum of the bladder is a congenital or acquired malformation of the urinary bladder. One of the bladder septums has sacciform formation. In children, it is congenital. Diverticulum can be single or multiple. Diverticulum is often located on the back or side septum of the bladder. The diverticulum is usually a stagnant urine. This may contribute to the development of cystitis. The child may complain of difficulty in urination, urination in two stages (the first emptying the bladder itself, and then diverticulum). The urologist can diagnose bladder diverticula based on cystoscopy and cystography. Treatment of this anomality is bladder surgery and its essence to eliminate the defect of the bladder septum.

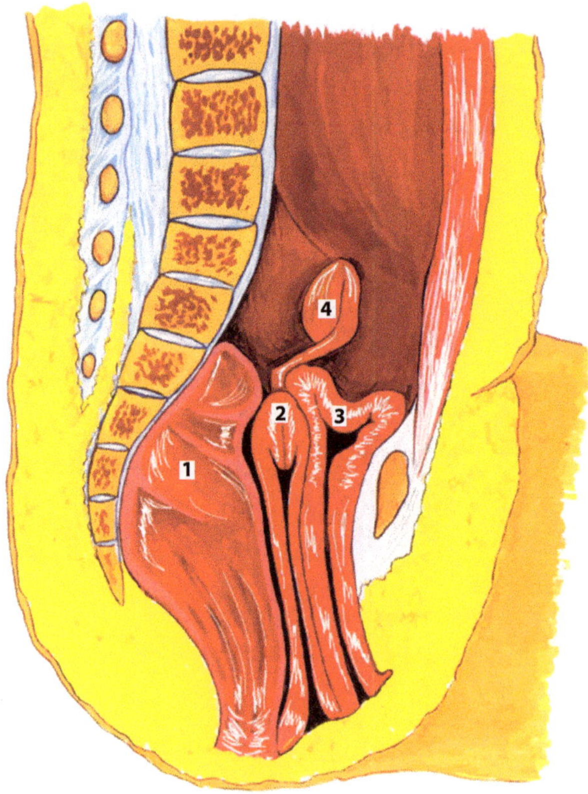

Figure 26 Sagittal section of pelvic girls 2 years old.
1 – rectum; 2 – uterus; 3 – vesica urinaria; 4 – ovarium sinister

Atresia of the rectum and anus, fistulas connecting the rectum with neighboring organs (Figure 28). Closing of the anus and rectum are the absence of the natural canals of the rectum and / or anus.

Classification of atresia: cloacal form, cystic, vaginal.

With Fistulas: in male urinary system (bladder and urethra), in the female reproductive system (uterus, vagina, vestibule), on the perineum in boys and girls (boys and on the scrotum and penis).

Without Fistula: the closure of the anus and rectum, the closure of the anal canal, covered the anus, the closure of the rectum during normal development of the anus.

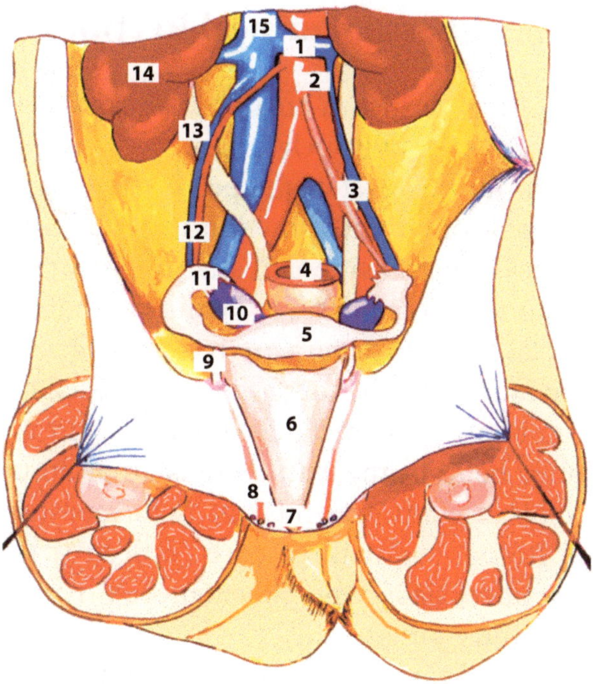

Figure 27 Pelvic organs of newborn girl.
1 – v. renalis sinistra; 2 – pars abdominalis aortae; 3 – vasa ovarica sinistra; 4 – rectum; 5 – uterus; 6 – vesica urinaria; 7 – urachus; 8 – a. umbilicalis; 9 – lig. teres uteri; 10 – ovarium; 11 – tuba uterine; 12 – vasa ovarica dextra; 13 – ureter; 14 – ren; 15 – v. cava inferior

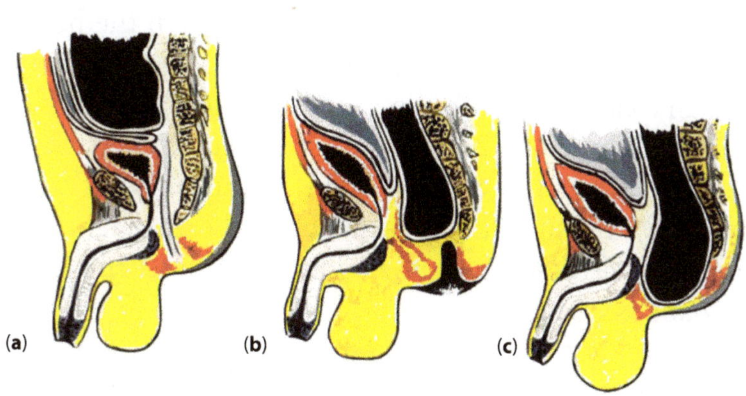

Figure 28 Rectal atresia.
a – atresia ani et recti; b – atresia recti; c – atresia ani

Anus. Ischiorectal Fossa. Perineal Rectum

The central body is the perineum anal or anal canal (canalis analis). The rectum consists of only of its pelvic part.

Anal canal begins at the level of the pelvic floor (m. Levator ani). There is an anal-rectal compound (junction anorectalis) that begins an anal transition zone (zona transitionalis analis). Its rectal epithelium is replaced by squamous epithelium type. On the external side of the anal canal the anal-rectal flexure (flexura anorectalis) corresponds to that place. In the formation of the bending the pubic-rectal muscle plays a big role (m. Levator ani). Anal-rectal flexure takes part in its support.

Around the anus beneath the skin there is an arbitrary outer anal sphincter (m. Sphincter ani externus). It is formed by the striated annular muscle. It consists of three parts: a deep (pars profunda), superfacial (pars superficialis) and subcutaneous (pars subcutanea). The canal ends the anus (anus).

The length of the anal canal varies depending on the individual and the physiological state of the anal sphincter, sex, age, weight and height and an average is of 3–4 cm. The anal canal has well marked longitudinal mucosal folds - anal or morganievy, poles (columnae anales). They are divided into anal sinuses (sinus anales). Their number varies from 5 to 13, and the depth is usually 3–4 mm. Bottom anal sinuses limit lunate flaps (valvulae anales). They are located 1.5–2 cm above the anus and form a comb-line (linea pectinata). On the border with the skin the anus-cutaneous line is determined (linea anocutanea). In front of the anal canal the bulbus penis, the posterior edge of the urogenital diaphragm and the center of the perineum, the back - coccyx and associated with it the canal septum corpus anococcygeum (lig. Anococcygeum) are located. On each side of the anal canal there are ischium-anal fossae, containing the adipose tissue (corpus adiposum fossae ischioanalis).

The blood supply to the anal canal is carried out by two or three lower rectal arteries (aa. Rectales inferiores) on each side. They pass through the fatty tissue of the sciatic-anal fossae in the frontal plane of the inner surface of the buttocks to the external sphincter.

Venous blood flows from the veins of the anal canal, which start from subcutaneous, submucosal and subfascialis venous plexus. Here blood flows from vv. rectales inferiores to vv. pudendae internae, further - through the veins of the rectum into the internal iliac vein and through the superior rectalis vein to the inferior mesenteric vein.

The innervation of the anal canal, the external sphincter and the skin around the anus carry inferior rectal nerves (nn. Rectales inferiores). They depart from n. pudendus in canalis pudendalis and after sciatic-anal hole direct to the rectum. The part of these nerves is sympathetic fibers that innervate the deep muscles of the canalis analis.

Subcutaneous lymph from the lymphatic network around the anus below valvulae anales occurs in the inguinal lymph nodes. From the posterior division of the anal canal in the area of attachment m. levator ani lymphatic vessels are directed to nodi lymphoidei pararectales.

Topography ischiorectal fossae. These are paired spaces, located on both sides of the canalis analis. Ischial-anal fossa has the shape of a quadrilateral pyramid. The top of it is directed upwards. The base of the pyramid is the layers of the soft tissues of the perineum: skin, subcutaneous fat and superficial fascia. Upper medial fossa septum is canalis analis and pelvic diaphragm.

The muscle goes obliquely from the side septum of the pelvis downwards and medially to m. sphincter ani externus. Between the muscle and the lateral septum an angle open downward is formed. At the angle obturator fascia and the upper and lower fascia of the pelvic floor (m. Levator ani) are connected forming the arcus tendineus m. levator ani. The lateral septum is formed of the lower half of the form m. obturatorius internus, fascia obturatoria and ischial tuberosity. Antertiorly the fossa limits the trailing edge of the deep transverse perineal muscles, posteriorly - the front surface of the gluteus maximus.

The splitting of the parietal fascia (fascia obturatoria) forms of pudendal canal (canalis pudendalis) or Olkok channel. It takes pudendal neurovascular bundle. A. et v. pudendae internae and n. pudendus enter the channel from a small obturator foramen and leaves it near the ischial tuberosity. The rectal arteries, veins and nerves go through the tissue of the ischial-anal fossa to the anus.

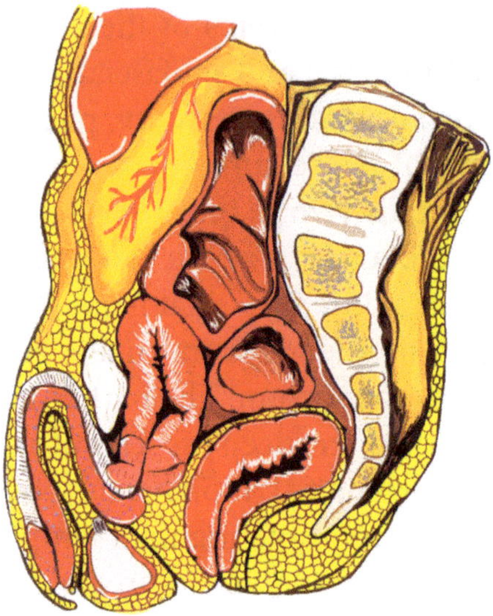

Figure 29 Sagittal section of pelvic of a boy, 5 years.

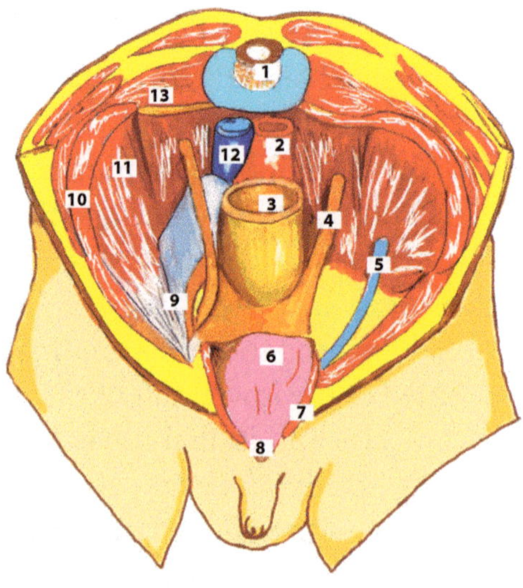

Figure 30 Pelvis organs if newborn boy.
1 – corpus vertebrae; 2 – aorta; 3 – rectum; 4 – ureter; 5 – v. spermatica interna; 6 – vesica urinaira; 7 – a.umbi; 8 – apex vesicae; 9 – anulus inguinalis profundus; 10 – m.iliacus; 11 – m.psoas major; 12 – v. cava inferior; 9 – m. gluteus maximus; 10 – cartilago coccyges; 11 – lig. anococcygeum; 12 – m.obturatorius internus; 13 – m. levator ani; 14 – tuber ischiadicum; 15 – centrum tendineum perinea; 16 – m. transversus perinea profundus

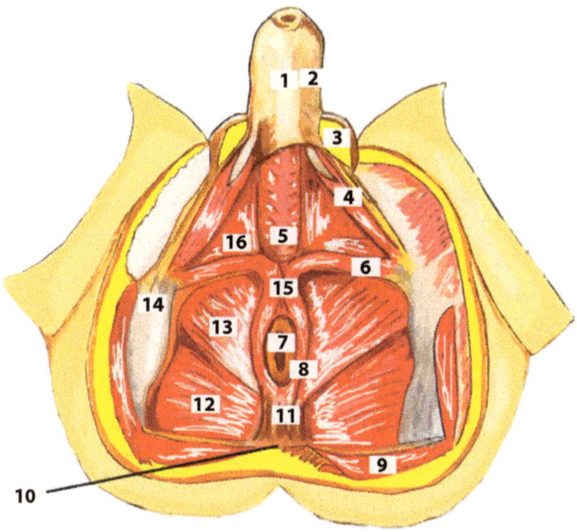

Figure 31 Topography of muscles and fascia of the perineum of newborn boy.
1 – corpus spongiosum penis; 2 – corpus; 3 – testis sinister; 4 ; m. ischiocavernosus; 5 – m. bulbospongiosus; 6 – m. transversus perinei superficialis; 7 – anus; 8 – m. sphincter ani externus; 9 – m. gluteus maximus; 10 – cartilago coccyges; 11 – lig. anococcygeum; 12 – m. obturatorius internus; 13 – m. levator ani; 14 – tuber ischiadicum; 15 – centrum tendineum perinea; 16 – m. transversus perinea profundus

Ischial-anal fossa is filled with fatty tissue (corpus adiposum fossae ischioanalis). Depth ischio-anal fossa from the skin surface to the top of the angle in adults is 5.0–7.5 cm. gradually it decreases anteriorly where it is 2.5 cm. In front of the fiber it is introduced into the gap between the m. levator ani and urogenital diaphragm (deep perineal pouch). Under the edge of the gluteus maximus, there is gluteal pocket (recessus glutealis). It binds fat sciatic-anal fossa deep space under the buttocks.

Perineal rectum. Crotch department is under the diaphragm of the pelvis and is the anal canal. The length of the crotch department is approximately 4 cm. It ends with the anus.

The anal canal has 8–10 regular longitudinal folds – bars (columnae anales) with recesses between them – anal sinuses (sinus anales). They end lunate folds – dampers (valvulae anales). Extends slightly zigzag line from the anal flap is called a gear and is the boundary between the ampoule and glandular epithelium squamous epithelium of the anal canal rectum. The annular space between the anal sinuses and anus called hemorrhoidal zone (zona hemorrhoidalis).

Submucosal layer is composed of loose connective tissue, which facilitates the easy displacement and stretching mucosa. The muscular septum has two layers: an inner and outer circular longitudinal. First one thickens at the top of the crotch department to 5–6 mm and forms an intrinsic sphincter (m. Sphincter ani int.). In the crotch portion of the colon longitudinal muscle, fibers are woven with the fibers of the muscles, lifting the anus (m. Levator ani), and partly with external pulp. The outer pulp (m. Sphincter ani ext.) consists of voluntary muscles, covering the crotch department and closing the rectum. It has a height of about 2 cm and a thickness of 8 mm.

Pelvic floor is formed muscle lifting the anus, and the PC muscle (m. Coccygeus) and fascia. Paired muscles, raising the anus, consist mainly of ilio-coccygeal (m. Iliococcygeus), pubococcygeal (m. Pubococcygeus) and pubo-rectal (m. Puborectalis) muscle and form a kind of funnel, lowered in a small basin.

The edges of the funnel are attached to the inner septums of the top department of a small basin. Below the center of the funnel is rectum. It is connected with the muscle fibers, lifting the anus.

By the edges of the stretched down oval abdominal covers attached its own fascia of the rectum, more dense behind and relatively less pronounced on the sides and the front turning into a dense aponeurosis predstatelnobryushinny (men) or rectovaginal aponeurosis (women). This aponeurosis is easily divided into two plates, one of which puts the prostate to the seminal vesicles, and the other - the anterior septum of the rectum; this facilitates the separation of these organs during surgery.

Lymphatic vessels and regional lymph nodes of the perineum.

The regional lymph nodes for the perineum are superficial inguinal lymph nodes located on top of the fascia lata, and deep inguinal lymph nodes localized underneath. Efferent vessels (vasa efferentia) deep lymph glanding in the groin sent to the pelvic cavity along a. iliaca externa, forming a dense network around the external iliac lymphatic plexus (plexus lymphaticus iliacus externus). They include 4–40 external iliac lymph glands (lymphoglandulae iliacae externae). The lymphatic vessels and glands of the lymph plexus are partly poured from the septums of the pelvis and the lower abdominal septum. The lymphatic vessels of the external iliac plexus flow into the lower lumbar lymph glands.

Surgical Anatomy of Congenital Malformations of the External Genitalia

Unfinished drooping of the testicle into the scrotum (cryptorchidism) is usually widespread in premature babies. It could be unilateral or bilateral. There are two variation of the malformation: intraperitoneal and inguinal. With age if treatment of such children is not carried out, then they complain of pain in the groin, increasing at weight lifting or physical activity. Sometimes eunuchiform type of physical development is developed in children (body hair on the female type, obesity) due to violation of the hormonal function of the testis. Physisians have to be able to distinguish the inguinal cryptorchidism from the false vagina.

Hydrocele and Edema of Spermatic Cord. Hydrocele is a manifestation of violation of obliteration of the vaginal process of the peritoneum. Unlike congenital inguinal hernia the communication of the lumen of the vaginal process of the peritoneum in the abdomen with edema is no more than 0.5 cm. Reason of accumulation of liquid is an imperfection of the lymphatic system in newborns and, in this regard, a delayed absorption. With age the possibility of completing of the obliteration of the vaginal process and the absorption properties of its membranes increase, which a significant part of children leads to self-cure of edema. Depending on the level of obliteration of the vaginal process of the peritoneum hydrocele, edema of spermatic cord, or cyst of the spermatic cord are developed. Clinically, testicular hydrocele is manifested by increase in size of half, while bilateral disease – all, due to the occurrence of scrotal elastic tumorous formation. When an isolated hydrocele, formation has a rounded shape, testicle is not palpable separately from this formation. When communicating hydrocele, spermatic cord formation has an oval shape, the upper edge of which extend to the external opening of the inguinal canal, making it difficult to examine. The testicle is palpated separately from this formation. Its sizes vary during the day, when the child's anxiety, it is increased. If the nature of the edema is the valvular, scrotum, increasing in size, becomes intense.

Phimosis is an inability to expose the glans. Only in 4% of newborn boys the foreskin is so mobile that it's possible to open the head of the penis. At the age of 6 months, the head of the penis is opened in 20% of boys and in 3 years old boys foreskin moves well and allows to expose the head of the penis in 90% of boys. Therefore, circumcision is not recommended for children up to 3 years old because it may cause an inability to expose the glans (phimosis).

Epispadias is a malformation of the urethra, characterized by dorsal dystopia of external opening of the urethra, nonclosure of the anterior septum and the splitting of the corpora cavernosa of the penis.

According to the classification epispadias of the penis is distinguished. It's more severe underdevelopment of the urethra, in which the splitting of the urethra extends over the entire organ. Its external hole opens at a different level of the dorsum of the penis, from the the head down to the radix.

There are complete epispadias – the most severe form of vice. The external urethral opening in the form of a wide funnel is located at the radix of the penis. In this form there is a complete splitting of the sphincter of the bladder and the urine constantly leaks out. The bladder is collapsed. Patients do not feel the urge to urinate.

Hypospadias. It is characterized by changes in distal part of the urethra. Clinically it's manifested be changes of the location of the external opening of the urethra.

Classification of Hypospadias

1. Hypospadias of penis:
   a. Hypospadias of corona glandis;
   b. Near-capitate, near-coronary hypospadias;
   c. Hypospadias of shaft of the penis;
   d. Penis-scrotal hypospadias;

2. Scrotal hypospadias;
3. The perinael hypospadias;
4. "Hypospadias without hypospadias".

Atresia of the vagina is isolated primary (congenital) or secondary (acquired) fusion of the vaginal septums. Clinically vaginal atresia is manifested by hematocolpos, hematometra, hematosalpinx, mucokolpos, the impossibility of sexual activity. Vaginal atresia is diagnosed during a gynecological examination, vaginal probe, ultrasound and MRI of the pelvis. Treatment of vaginal atresia is only surgical – emptying the hematocolpox and plastic of the fully functioanal vagina. Complications of the pathology are the ascending infection, peritonitis, sepsis, repeated fusion of the vaginal septums.

Pathotophography of the Peivis

Uterine fibroids are hormone-dependent, benign, tumor-like the formation of the uterus, coming from its smooth muscle and connective (fibromyoma) tissue. Uterine fibroids are single, but more often - in the form of multiple myomatous nodes with different localization. Uterine fibroids can range in size from a small nodule to a tumor weighing about a kilogram, when it is easily determined by palpation of the abdomen.

Pathotography (Figure 32). With subserous uterine myoma, the node develops on a broad base or long leg. Localized subserous fibroids subperitoneally, on the surface of the uterus under the serous membrane. Submucous (submucous) uterine myoma grows in the uterine cavity. With interstitial myoma of the uterus tumor site is located in the thickness of the muscular wall of the uterus. The diagnosis of uterine fibroids can be established already at the primary gynecological examination. With a two-handed vaginal examination, a dense, palpable enlarged uterus with knobby, knobby surface. More reliably determine the size of uterine fibroids, its localization and classification allows ultrasound examination of the pelvic organs.

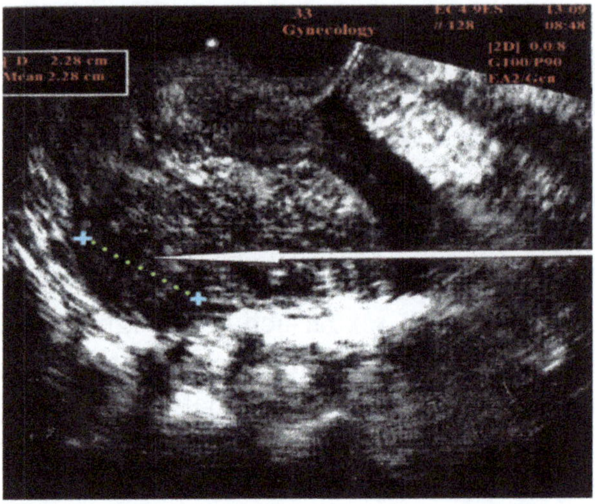

Figure 32 Uterine fibroids.

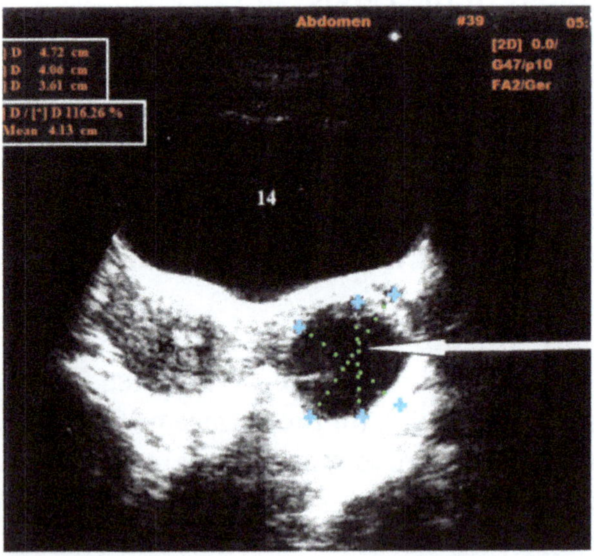

Figure 33 Cyst of the left ovary.

Ovarian cyst - tumor-like formation of an ovary of benign nature. It is a cavity on the leg, filled with liquid contents and tending to increase in size due to the accumulation of a secret.

Pathotography (Figure 33). There are ovarian cysts (follicular, cysts of the corpus luteum, endometrioid, etc.) and supraovarial appendage (paraovarial). They are often asymptomatic, can manifest discomfort and pain in the lower abdomen, menstrual dysfunction, dysuric disorders (with pressure on the bladder). Legs of a cyst, rupture of a capsule, a picture of an acute abdomen and peritonitis are complicated by torsion of the cyst leg.

Depending on the nature of the origin of the formation and its contents, the following cysts are distinguished: follicular; cyst of the corpus luteum; paraovarial; endometrioid; dermoid; mucinous.

Follicular cysts and cysts of the corpus luteum are functional formations that form in the ovary tissue itself and are directly related to the cyclic changes occurring in it. Follicular cyst formation occurs at the site of the unexploded follicle, and the cyst of the corpus luteum is on the site of the unregressed follicle yellow body. Pathological cavities in these types of ovarian cysts are formed from the shells of the follicle and the corpus luteum, respectively. The basis of their occurrence are hormonal disorders. Usually, a follicular ovarian cyst and a corpus luteum cyst do not reach. They are of considerable size and can disappear on their own as the secret and decay of the cystic cavity dissolve into them.

Paraovarian cysts originate from the supraovarial appendages without involvement of the ovarian tissue in the process. Such ovarian cysts can reach gigantic proportions. Endometriotic cysts are formed from particles of the uterine mucosa (endometrium) with its pathological focal growth on the ovaries and other organs (endometriosis). The content of endometrioid ovarian cysts is old blood. without involvement of the ovarian tissue in the process. Such ovarian cysts can reach gigantic proportions. Endometriotic cysts are formed from particles of the uterine mucosa (endometrium) with its pathological focal growth on the ovaries and other organs (endometriosis). The content of endometrioid ovarian cysts is old blood.

Ovarian mucinous cysts are often multichamber, they are filled with thick mucus (mucin), which is produced by the cyst's internal mucous membrane. Endometrioid and mucinous ovarian cysts are more prone

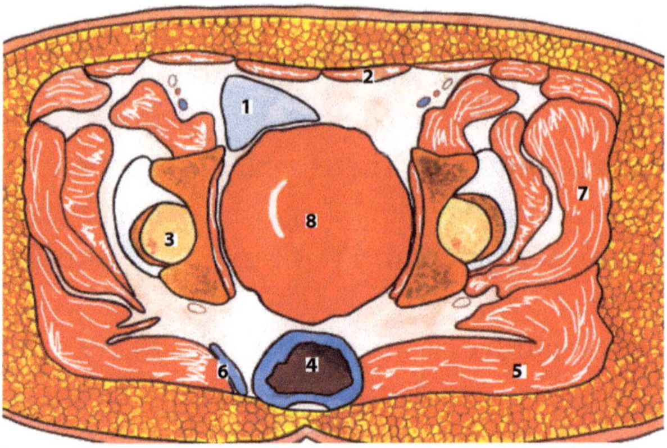

Figure 34 Cross section of a female small pelvis.
1 – vesica urinaria; 2 – m. rectus abdominis; 3 – acetabulum; 4 – rectum; 5 – m. gluteus maximus; 6 – lig. sacrospinale; 7 – m. gluteus maximus; 8 – leukosarcoma of the uterus

to degeneration into malignant neoplasms. Congenital ovarian cysts include dermoid cysts that form from embryonic buds. They contain fat, hair, bones, cartilage, teeth and other fragments of body tissues.

Echographic detection of ovarian cysts is based on the diagnosis of the liquid contents of cystic formation. So, the main echographic signs ovarian cysts are large, the presence of internal partitions, parietal seals, heterogeneous fine internal echostructure.

Leiomyosarcoma (Figures 34–36) (malignant leiomyoma) is an immature malignant tumor of smooth muscle. Most often localized in the gastrointestinal tract, mainly in the colon, then retroperitoneal, in the soft tissues of the limbs, in the uterus. The development is characterized by a significant diversity in relation to the number and size of tumor nodes, which can be located in its various sections, including in the area of the cervix and the isthmus of the uterus. With the development of multiple nodes or significant tumor size, the uterus deformity occurs.

The bladder in the female pelvis lies in the pelvic cavity deeper than in men. In the peritoneal floor behind the bladder there is a deformed body of the uterus and intestinal loops entering the excavatio vesicouterine. The deformed body of the uterus does not significantly press down the bladder.

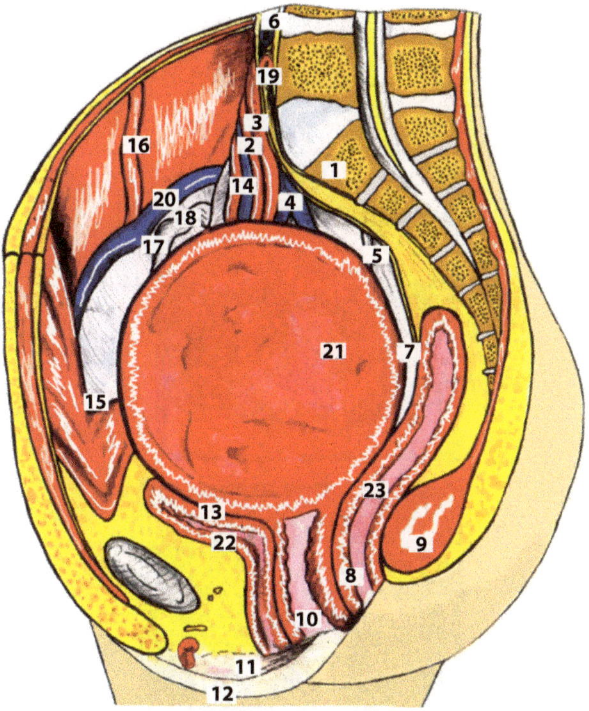

Figure 35 Sagittal cutting female pelvis.
1 – cape; 2 – ureter dexter; 3 – a. iliaca interna; 4 – v. iliaca interna; 5 – plica rectouterinalis; 6 – v. coniunnis iliaca sinistra; 7 – excavation rectouterinalis; 8 – m. sphincter ani inicrnus; 9 – m. sphincter ani extemus; 10 – vagina; 11 – labium minus pudendi; 12 – labium majus pildellda; 13 – excavatio vesicouterinalis; 14 – vasa ovarii (in supporting ligament); 15 – lig. teres uteri; 16 – a. iliaca externa; 17 – tuba uterine; 18 – ovarium; 19 – a. iliaca comunnis dextra; 20 – v. iliaca externa; 21 – leucosarcoma of uterus; 22 – vesica urinaria; 23 – rectum

The uterus is located in the pelvis between the bladder in front and the back of the rectum. It consists of two sections: the upper - body, corpus, bottom, fundus, and lower - cervix, cervix uteri. The body of the uterus in the bottom area is slightly increased due to tumor formation. The tumor site has a spherical the 2×2 cm shape, the characteristic whitish surface, is clearly separated from the surrounding myometrium. The side edges of the uterus, margo uteri dexter et sinister, are slightly flattened. Most of the uterus is located in the upper, peritoneal floor of the pelvis.

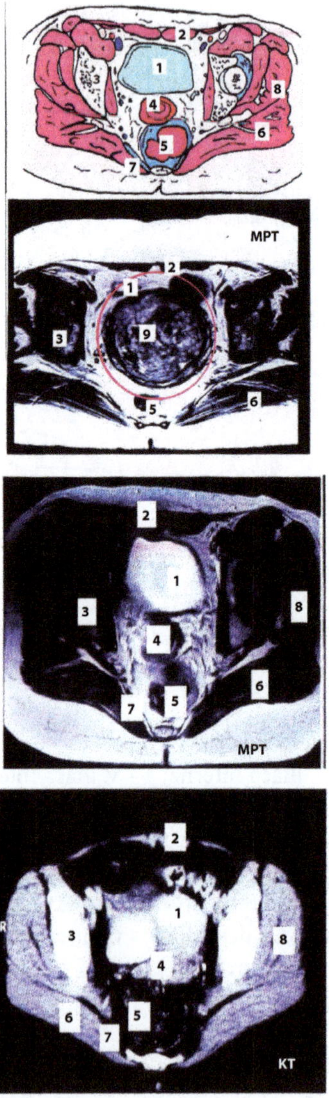

Figure 36 Cross section of the female pelvic organs.
1 – vesica urinaria; 2 – m. rectus abdominis; 3 – acetabulum; 4 – uterus; 5 – rectum; 6 – m. gluteus maximus; 7 – lig. sacrospinale; 8 – m. gluteus medius; 9 – uterine leiomyosarcomaa
– In MRI and CT syntopy of organs are visualized – from the anterior abdominal septum to the spine in the midline are located rectus abdominis (2), the bladder (1), the uterus (4), rectum (5) are located.
b – In MRI uterine leiomyosarcoma (9) is visible and it occupies a significant part of the pelvic cavity. Bladder (1) is offset laterally to the midline due to compression of the pathological structure. The rectum (5) is also displaced laterally from the midline.

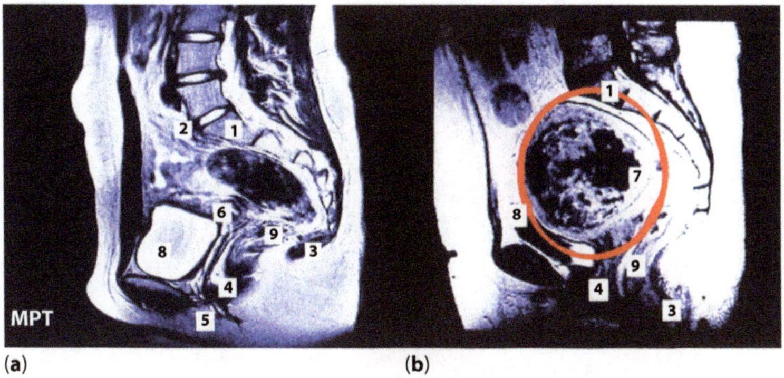

Figure 37 Sagittal section of the female pelvic organs. MRI.
1 – cape; 2 – ureter dexter; 3 – m. sphincter ani internus; 4 – vagina; 5 – labium minus pudenda; 6 – excavatio vesicouterinalis; 7 – uterine leiomyosarcoma; 8 – vesica urinaria; 9 – rectum

Blood supply is carried out by two uterine arteries, aa. uterinae, ovarian arteries, aa. ovaricae (from the abdominal aorta) and arteries of the round ligament of the uterus, aa. lig. teretis uteri.A. uterina is a branch of the internal iliac artery that supplies blood to the bottom of the uterus. A. uterina is slightly dilated due to increased blood supply to the tumor site. Possible vascularization (the formation of blood vessels in the node).

Venous outflow from the uterus occurs first in the uterine venous plexus, plexus venosus uterinus. It widely anastomizes with all the veins of the pelvis, but primarily with the venous plexus of the vagina, plexus venosus vaginalis. From the plexus blood through the uterine veins flows into the internal iliac veins. Outflow from the bottom of the uterus, ovaries and tubes occurs in the inferior vena cava through vv. ovaricae, which are also slightly extended.

Lymph drainage. The bottom of the uterus is moderately swollen, due to increased formation of lymph, lymph nodes in the area of the bottom of the uterus are also enlarged. From the bottom of the uterus through the diverting lymphatic vessels of the round ligament of the uterus, ligg. teres uteri, lymph partially flowing to the inguinal lymph nodes.

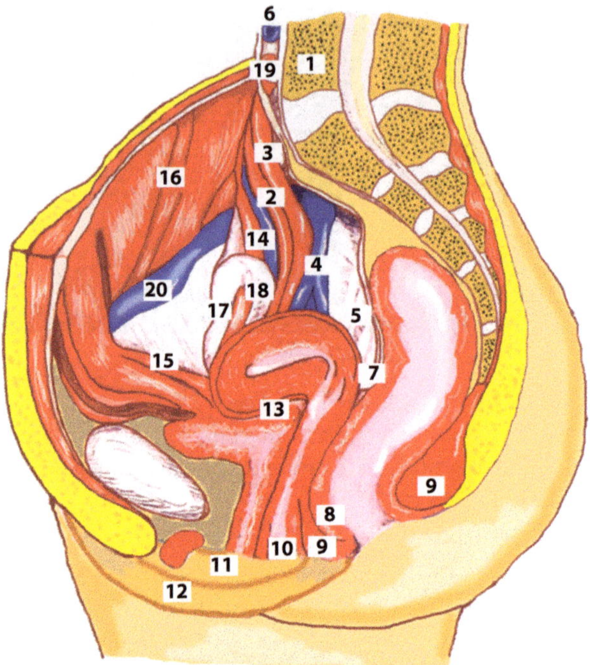

Figure 38 Interrelation of the organs of the female pelvis.
1 – cape; 2 – ureter dexter (relief); 3 – a. iliaca interna; 4 – v. iliaca interna; 5 – pl. carectouterinalis; 6 – v. communis iliaca sinistra; 7 – excavatio rectouterine; 8 – m. sphincter ani internus; 9 – m. sphincter ani externus; 10 – vagina; 11 – labium minus pudenda; 12 – labium majus pudenda; 13 – excavatio vesicouterina; 14 – vasa ovarii (in supporting ligament); 15 – lig. teres uteri; 16 – a. iliaca externa; 17 – tuba uterinae; 18 – ovarium; 19 – a. iliaca communis dextra; 20 – v. iliaca externa

Characteristic is the diffuse expansion of the uterus. The edges of the tumor are less clearly delineated than in the case of leiomyoma (uterine leiomyosarcoma is usually impossible to separate from the myometrium). On the incision, the tumor is edematous, fleshy, with local necrosis and hemorrhage, without the typical concentric structure inherent in leiomyoma.

In comparison with the normal tomogram of the patient with uterine leiomyosarcoma (21) the MRI shows a significant compression of the bladder (22), tumor flattening the rectum (23) and the vagina (10).

Uterine myxoid leiomyosarcomas are characterized by a gelatin-like appearance and relatively well-defined boundaries. Some tumors can grow into a wide ligament of the uterus.

Based on a certain pattern of damage to individual groups of lymph nodes, there are three stages of metastasis:

1. in the external, internal iliac and obturator nodes;
2. to the common iliac nodes;
3. in the lumbar nodes.

In most cases, a known stage of lesion of the lymphatic apparatus in cervical cancer remains. However, the diversity of lymphatic drainage pathways, wide anastomoses of individual groups of lymph nodes with each other, the constant functioning of the lymphatic vessels and nodes often violate this pattern and predetermine the variability of lymphogenous localization metastasis of cervical cancer.

The external and internal iliac and obturator lymph nodes are most often affected, which for the female genital organs are regional lymph nodes of the second stage. Revision of the lumbar lymph nodes during surgery for cervical cancer is required.

Lymph from the body of the uterus is collected in the diverting lymphatic collectors, reaching in the parametric space through the peripheral nodes to the internal and external iliac nodes, as well as posteriorly, to the sacral nodes and the promontorium nodes.

The MRI (Figure 36 and 37) shows uterine leiomyosarcoma, which occupies a significant part of the pelvic cavity. The bladder is displaced laterally relative to the median line due to compression of the pathological formation. The rectum is also displaced laterally relative to the midline.

Liquid formation (Figures 39 and 40) is a cavity filled with liquid contents delimited by a capsule and tissue from the main body of the organ. They have a various form, more often - rounded, as a rule, clearly defined objects. Often, in the lumen of liquid formations, various structures are revealed — partitions, inclusions, amorphous masses.

Retroperitoneal fluid masses are inflammatory infiltrates, cysts, abscesses, hematomas in the lysis stage, cystadenomas, cystadenocarcinomas. The source of these pathological processes are mainly

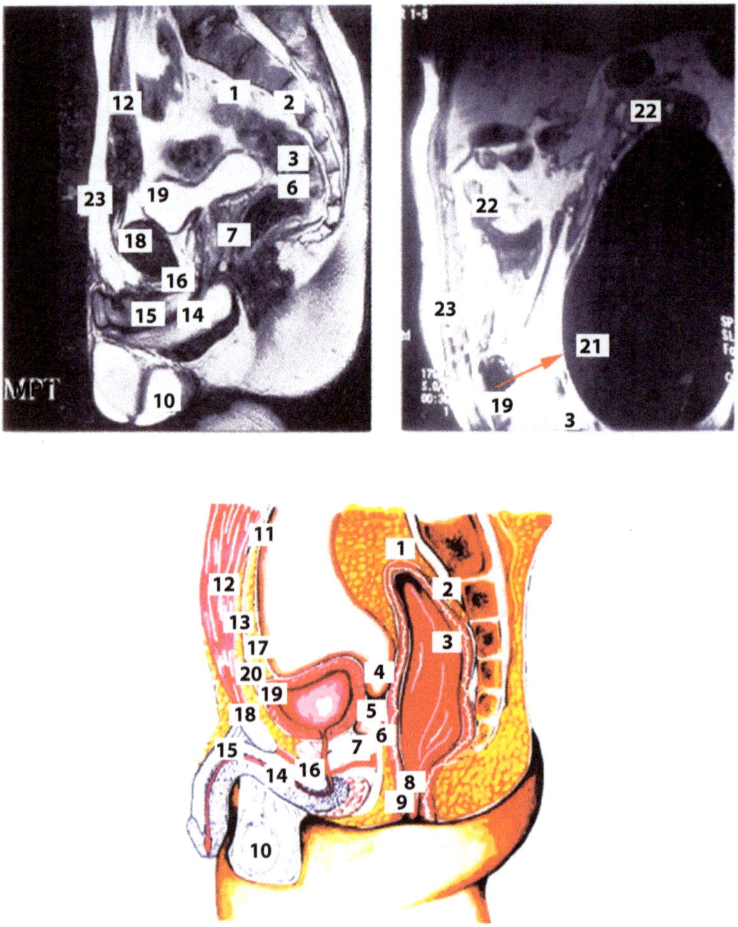

Figure 39 Sagittal section of the male pelvic organs.
1 – panniculus parietalis; 2; 3 – rectum; 4 – excavatio rectovesicalis; 5 – vesicula seminalis; 6; 7 – prostata; 8 – m. sphincter ani internus; 9 – m. sphincter ani externus; 10 – testis; 11 – peritoneum; 12 – m. rectus abdominis et fascia transversalis; 13 – fascia prevesicalis; 14 – urethra; 15 – corpus cavernosum penis; 16 – diaphragm urogenitalis; 17 – paniculus preperitonealis; 18 – symphysis; 19 – vesica urinaria; 20 – paniculus prevesicalis; 21 – fluid accumulation in the retroperitonealis space; 22 – lemniscus intestine; 23 – paries abdominalis anterior.
a – the sagittal section of the male pelvis in the norm. In the MR-tomogram not all the anatomical structures indicated in the signatures to the figure are visible. On the MR-tomogram there is no the recto-vesical pouch (4), since it is a slit-shaped space undetectable in norm on MRI.
b – the sagittal section of the male pelvis. Fluid accumulation.

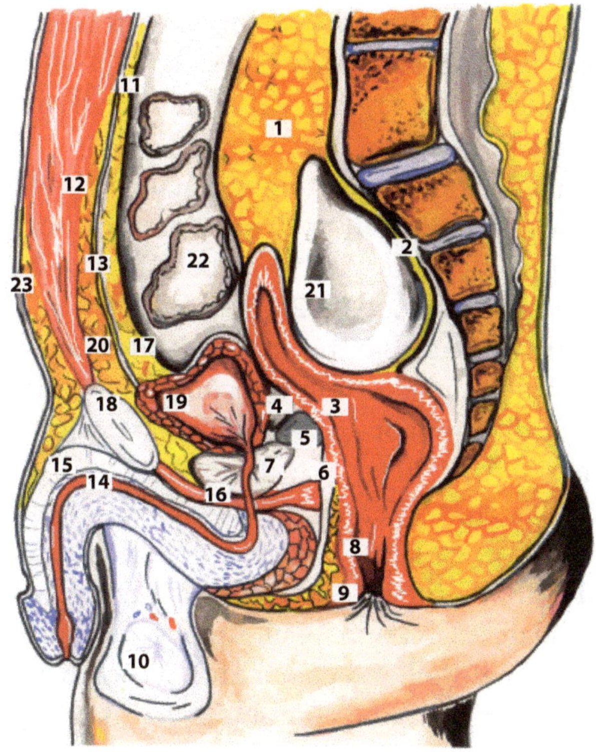

Figure 40 Sagittal cut male pelvis.
1 – panniculus pararectalis; 2 – parietal sheet of pelvic fascia; 3 – rectum; 4 – recessus rectovesicale; 5 – vesicula seminalis; 6 – aponeurosis peritoneo perineus; 7 – prostata; 8 – m. sphincter ani internus; 9 – m. sphincter ani externus; 10 – testis; 11 – peritoneum; 12 – m. rectus abdominis et fascia transversus; 13 – fascia precystica; 14 – urethra; 15 – corpus cavernosum penis; 16 – diaphragma urogenitale; 17 – panniculus preperitonealis; 18 – symphysis; 19 – vesica urinaria; 20 – panniculus precysticus; 21 – fluid formation in the retroperitoneal space; 22 – lemniscus intestini; 23 – paries abdominalis anterior

the organs of the retroperitoneal space - the pancreas, the descending and horizontal parts of the duodenum, the ascending and descending colon, adrenal glands, kidneys, ureters, large vessels, nerves, lymphatic vessels and fatty tissue.

Complications are usually associated with the nature of the pathological process. They can manifest as compression of surrounding organs (cysts, hematomas, cystadenocarcinomas); the appearance of areas of inflammation in healthy tissues (abscesses); spread to neighboring areas, the development of secondary osteomyelitis (abscesses, cystadenocarcinomas); bleeding, perforation (cysts).

In the MR-tomogram the organs of the abdominal cavity and small pelvis are in the sagittal section.

In comparison with the normal tomogram, a gross violation of the topography of the abdominal and pelvic cavities, caused by the accumulation of fluid in the recto-vesical space (21), can be seen in the sagittal section. The bowel loops (22), the bladder (19), the rectum – all these organs are displaced to the anterior abdominal wall (23) and compressed, as can be judged from the considerable deformation of the contours in the direction of flattening and narrowing of the lumens of hollow organs.

The figure shows a gross violation of the topography of the abdominal cavity and small pelvis, caused by the accumulation of a significant amount of fluid in the retroperitoneal space at the level of the sacrum. Intestinal loops, bladder, rectum — all these organs are displaced to the anterior abdominal wall and constricted, as can be seen from the considerable deformation of the contours towards the flattening and narrowing of the lumens of hollow organs.

Prostate cancer (Figures 41 and 42) develops in the glandular elements of the parenchyma and the terminal tubules of the peripheral zone of the gland adjacent to it. In the process of aging in the peripheral zone, the number of lesions of atypical hyperplasia, which differs from glandular hyperplasia associated with the development of prostate adenoma and the disorganization of glandular structures, is increasing. Atypical hyperplasia develops in areas with preserved secretory activity of the epithelium, which subsequently undergoes diffuse glandular hyperplasia, which reflects the biological relationship between these processes and the emergence of the histological picture of cancer. The evolution of prostate cancer includes a prolonged precancerous condition in which the increased sensitivity of the normal epithelium to androgenic stimulation and the consequent loss of hormonal dependence are key factors in the stimulation of malignant growth. The atypicalness of the cells that appears reflects the loss of their androgen dependence.

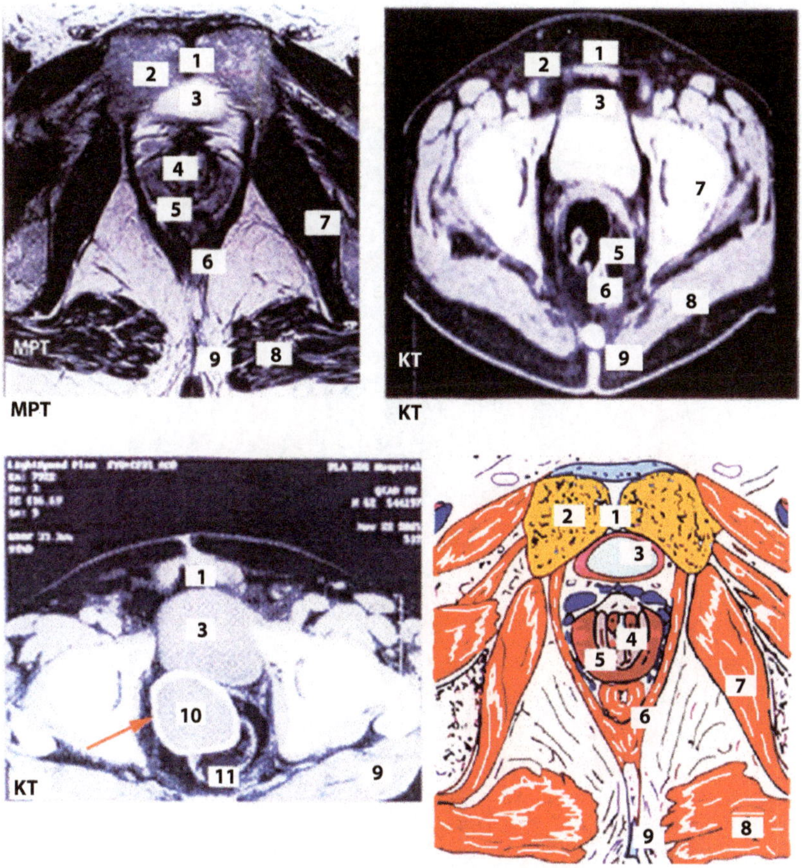

Figure 41 Cross section of the male pelvic organs.
1 – symphysis pubis; 2 – os pubis; 3 – vesica urinaria; 4 – urethra; 5 – prostata; 6 – anus; 7 – m. obturatorius internus; 8 – m. gluteus maximus; 9 – lig. anococcygeum; 10 – cancer prostatae; 11– rectum.
a – At MRI and CT syntopy of organs are visualized – from the anterior abdominal wall to the spine in the midline the pubic symphysis (1), the bladder (3), the urethra (4), the prostate (5), the anus (6) the anal-coccygeal ligament (9) are located.
b – In CT scan prostate cancer (10), compressing the rectum (11) and the urethra (4) and the bladder (3) are visible.

In the initial stage of the development of prostate cancer, a microscopic focus of malignancy is determined, including the area of significant or mild atypical proliferation of the epithelium. Most of these foci are localized in the peripheral part of the gland and measure 0.2–2.1 mm. Histologically, they relate to clear cell-tubular-alveolar

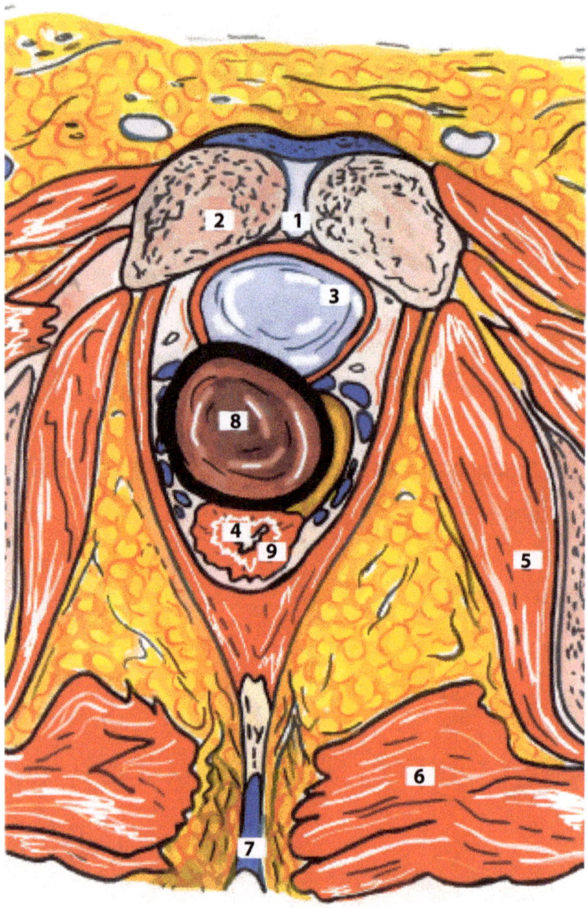

Figure 42 Cross section of the male pelvis.
1 – simphysis pubiti; 2 – os pubis; 3 – vesica urinaria; 4 – anus; 5 – m. obturatorius internus; 6 – m. gluteus maximus; 7 – lig. anococcygeum; 8 – cancer prostatae; 9 – rectum

or dark-cell tubular carcinoma. With the development of a malignant process, numerous microscopic infiltrates spread among normal glandular tissue. At the same time, the site of malignancy often proves to be ulticentric with the growth of connective tissue. Gradually, the condensation of the pathological focus into a cancer node palpable during rectal examination.

Localization of ulcers in the cellular tissue spaces of the pelvis (Figure 43). In the pre-cellular fiber space there are loose fatty tissue, vesical and prostatic venous plexuses and arteries involved in the

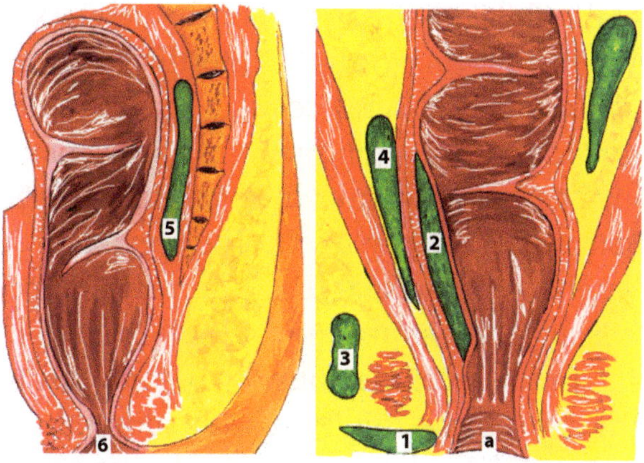

Figure 43 Localization of ulcers in the cellular spaces of the pelvis: anteroposterior (a) and side view (b).
1 – subcutaneus; 2 – submucosa; 3 – ishiorectalis; 4 – pelvirectalis; 5 – retrorectalis

blood supply to the bladder. In case of fractures of the pubic bones, the pre-vesicular cell space serves as a place for the accumulation of spilled blood, it is also possible that the bladder is wounded to form urinary streaks with the subsequent formation of cellulose phlegmon. Purulent leakages from the pre-vesicular space can spread into the lateral parietal spaces of the pelvis, into the vesicle and visceral fat, into the free abdominal cavity, into the vagina of the rectus abdominis, into the navel, under the skin.

The prostate gland, the end sections of the ureters, the vas deferens with their ampoules, the seminal vesicles, loose cellulose and the prostate venous plexus are located in the posterior hypocellular tissue space. It is possible the spread of purulent leakage from the posterior vesicle cellular space into the posterior visceral space of the bladder; in the area of the inguinal canal along the vas deferens; in the retroperitoneal tissue space along the ureters; in the urethra; into the rectum.

In the fatty tissue of the posterior rectum, the superior rectal artery, the median and lateral branches of the sacral artery are located; inferior hypogastric (pelvic) plexus; sacral plexus; sacral group of lymph nodes. The spread of purulent outflows from the posteriorly rectal intestinal space is possible in the retroperitoneal cellular tissue space, the lateral parietal cellular tissue spaces of the pelvis, visceral cellular tissue space of the rectum (between the intestinal wall and its fascia).

In the fatty tissue of the lateral cellular spaces, blood vessels are located (common and internal iliac arteries and veins); lymph nodes and vessels along the common and internal iliac arteries; nerve visceral plexus; ureters, ejaculatory ducts. The spread of purulent leakages from the lateral cellular spaces of the pelvis is possible in the gluteal region through the supra- and subglobular foramen; in the posteriorly rectal and pre-cellular fibrous spaces; through the obturator canal to the area of the thigh muscles; along the vessels in the visceral tissue space of the pelvic organs; in the retroperitoneal tissue space.

Between the visceral fascia of the pelvis and internal organs is located fatty tissue. There are visceral cellulose spaces around the bladder, prostate gland, seminal vesicles, rectum in men and the bladder, uterus, vagina, rectum in women.

In the circulatory fiber (parametrium) distinguish the front parametrium that separates the cervix from the bladder, and the rear, separating the vagina from the rectum.

In the circulatory fiber (parametrium) distinguish the front parametrium that separates the cervix from the bladder, and the rear, separating the vagina from the rectum. The lateral parametrium is located at the base of the wide ligament of the uterus between its sheets and passes directly into the parietal fiber of the pelvis.

The circulatory tissue space contains the uterine artery and the ureter intersecting it; ovarian vessels; uterine venous plexus and utero-vaginal nerve plexus. Purulent flows from the circulatory cellular space spread along the ureters and ovarian vessels to the retroperitoneal tissue; in the fiber of the ileal fossa; in the tissue of the gluteal region; in the inguinal canal.

The near-bladder cellulose space is widely communicated with the preperitoneal space, and its lower divisions - with the lateral wall spaces.

Paraprostatic space is limited to the fascial capsule of the prostate gland. Pustules from the prostate gland can break into the urethra, the rectum, anteriorly - in the pre-bladder space, sciatic-rectal fossa, and even through the peritoneal perineal fascia into the peri-rectal fiber.

The seminal vesicles are surrounded by a fascial sheath. The near-bladder fiber cell space continues along the vas deferens up to the deep ring of the inguinal canal, along the ureter up and down it communicates with the paraurethral cell space.

The peri-rectal cellular tissue space is limited by the fascial rectal case, which is not closed from the sides. Visceral fiber spaces communicate freely with other cellular tissue spaces of the pelvis, pus can spread in them with the formation of pelvicorectal, ischiorectal, retrorectal phlegmon and abscesses.

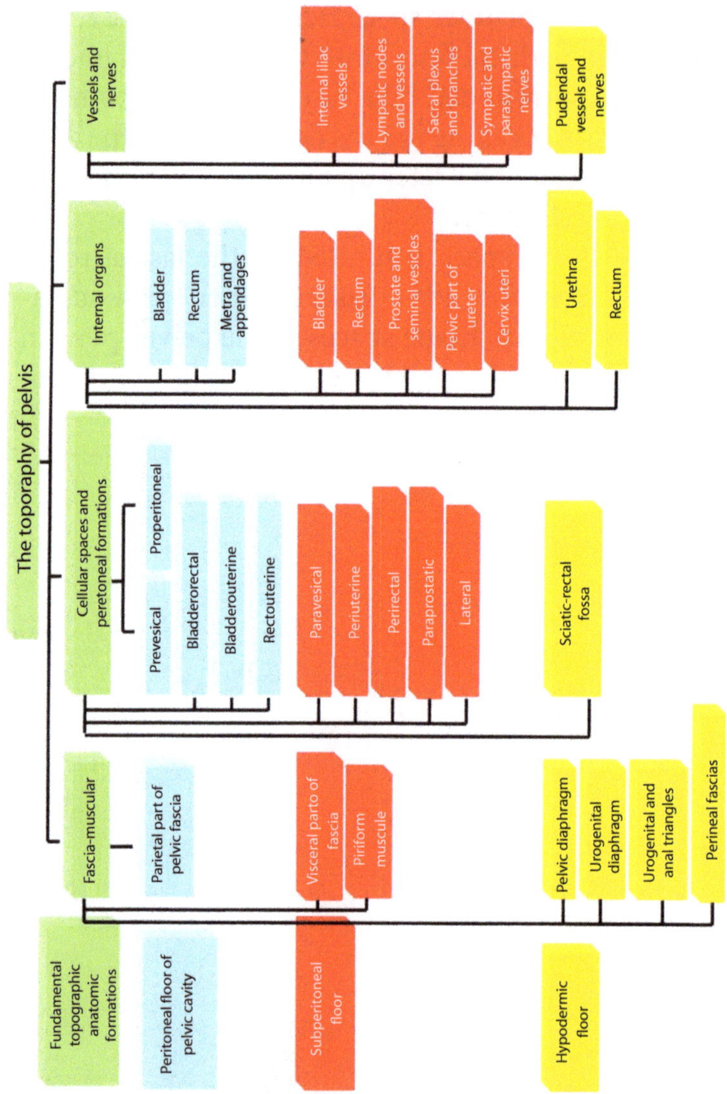
The toporaphy of pelvis
Vessels and nerves
Internal iliac vessels
Lympatic nodes and vessels
Sacral plexus and branches
Sympatic and parasympatic nerves
Pudendal vessels and nerves
Internal organs
Bladder
Rectum
Metra and appendages
Bladder
Rectum
Prostate and seminal vesicles
Pelvic part of ureter
Cervix uteri
Urethra
Rectum
Cellular spaces and peretoneal formations
Prevesical
Properitoneal
Bladderorectal
Bladderouterine
Rectouterine
Paravesical
Periuterine
Perirectal
Paraprostatic
Lateral
Sciatic-rectal fossa
Fascia-muscular
Parietal part of pelvic fascia
Visceral parto of fascia
Piriform muscule
Pelvic diaphragm
Urogenital diaphragm
Urogenital and anal triangles
Perineal fascias
Fundamental topographic anatomic formations
Peritoneal floor of pelvic cavity
Subperitoneal floor
Hypodermic floor

Part 2: The Spine

Topographic Anatomy of the Spine

Parts, external reference points. There are 5 sections of the spine, called regions: the cervical (7 vertebrae, C1-C7); thoracic (12 vertebrae, Th1-Th12); lumbar (5 vertebrae, L1-L5); sacral (5 vertebrae, S1-S5); coccygeal portion (3–5 vertebrae, Co1-Co5). Figure 32 – MR-visible skeletopy, CT scan of the head and neck: the tongue (6) is located at I-III cervical vertebrae, pharynx (9) – at the level of IV-V cervical vertebrae, the cervical part of the esophagus (10) – at the level of V-VII cervical vertebrae, larynx (7), – at III-IV of the cervical vertebrae, trachea (8) begins at the level of the vertebra V). Figure 33 – MR-CT scan shows all the anatomical structures. Visible skeletopy of thoracic organs: thyroid (1) is located 2 cm above the breastbone (10) at the level III-VI thoracic vertebra, disposed heart (6, 7), liver (5) are located at the level of 7–12 right ribs.

There are 2 types of vertebral curves: lordosis and kyphosis. Ventral spinal lordoses (frontward) are cervical and lumbar (Figure 34). Dorsal spinal kyphosis (back) are thoracic and sacral. Curvature of the spine helps a human to keep his balance. When walking, jumping and running it compresses like a spring and softens shocks. The spine is involved in the formation of the posterior septum of the chest and abdominal cavity, pelvis. The spine is the place place of origin and attachment of muscles of the trunk and limbs.

The spine and the spinal canal. The spine (columna vertebralis) is the axial skeleton. It consists of 32–33 vertebrae (7 cervical, 12 thoracic, 5 lumbar, 5 sacral, connected to the sacrum, and the coccygeal 3–4). Between the vertebrae there are 23 intervertebral discs. The typical vertebra has the corpus, opening and closing vertebral arch, two transverse and two superior and two inferior articular processes. The vertebral corpuses are consisted of spongy substance. Outside they are covered with a layer of compact substance. The vertebrae vary in shape, the curvature of the superior and inferior surfaces, height of anterior and posterior divisions. The spinous processes are different from each other in size, shape and orientation in space. They limit the extension of the spine, especially in the thoracic region (Figure 35). The transverse processes of the vertebrae, except for infants, include the rudiments of ribs. The transverse processes of the cervical vertebrae (Figure 20) contain holes. These holes constitute a channel for vertebrate blood vessels and nerves. The spatial orientation of the articular processes of the vertebrae is not the same variety. I and II of the cervical vertebrae (atlas and axis) have a special form. Due to the structure of the connections between the occipital bone, atlas and axial vertebra provide mobility of the head in three dimensions. The vertebrae are linked by all the main types of the joints. In adults, the corpus of the vertebrae, except for sacral, are separated by intervertebral discs. These discs are composed of fibrous ring and nucleus pulposus. Collagen fibers form the peripheral portion of the fibrous ring. They have a spiral intersecting direction. The inner part of the ring consists of fibrocartilage. The fibers of the fibrous ring are joined to the surface of the vertebral bodies, anterior and posterior longitudinal ligament. In the thoracic region the fibrous ring is connected with joints of the heads of the ribs. Corpuses of vertebrae are separated by the intervertebral discs with a thin layer of hyaline cartilage. Plastic nucleus pulposus contains an amorphous intercellular substance, which is diluted in the central part. A few vertebrae are displaced posteriorly in the inferior thoracic and lumbar spine sections. It serves as an elastic shock absorber. The intervertebral discs make up about 1/4-1/5 of the length of the spine in adult. The arcs of each pair of the vertebrae are connected by two flat joints, ligaments and white ligamentous apparatus developed between the spinous transverse processes. There are vertebrate's clippings on the arcs of the vertebrae. They make up the intervertebral foramens. The spinal nerves and blood vessels goes through these foramens.

Individual and Age Differences of the Spine

The spine of the child at birth is absolutely straight. Only years after it takes an S-shape. It happens gradually. Firstly cervical lordosis of the baby's spine is formed. It is a natural bend forward. It is formed when the child lies on his stomach and starts to raise the head. This is about 4 months of age. By 7 months thoracic kyphosis is formed. It is a natural bend backward. It occurs when a child begins to sit. Last physiological bend is formed by 8–9 months when lumbar lordosis is formd (Figure 41). At this time, a child learns to stand on the legs.

The length of the spine in adult women is 60–65 cm. In men, it ranged from 60 to 75 cm. In old age the length of the spine is reduced by about 5 cm due to the increase in the age of the spine bends and reducing the thickness of the intervertebral discs. The width of the vertebrae is reduced from the bottom to the top. At the level of XII thoracic vertebra it is 5 cm. The greatest transverse dimension (11–12 cm) of the spine is at the base of the sacrum.

The Spinal Cord and Nerve Roots

The spinal cord is a section of the central nervous system. It consists of millions of nerve cells and nerve fibers. The spinal cord is surrounded by three meninges (pia mater, dura mater and arachnoid mater), and is located in the spinal canal.

The dura mater forms a sealed bag of connective tissue (dural sac). In this sac spinal cord and a few centimeters of the nerve roots are located. The spinal cord in the dural sac (Figure 42) restricts the cerebrospinal fluid.

Spinal cord starts from the brain and ends between the first and second lumbar vertebrae. Roots depart from the spinal cord nerve. They form the so-called cauda equine below the end of the spinal cord. The roots of the cauda equine are involved in the innervation of the lower limbs. Peripheral nerves form plexuses: the cervical, brachial (innervate the chest and abdomen), lumbar and sacral (innervate the legs, crotch and pelvic organs (bladder, rectum)) (Figure 39). Doctor, determining disorders of sensation or motor function of which area of the body are manifested, can assume the level of the damage of the spinal cord (Figure 40 – MR-CT scan visualizes spinal cord injury (1) as a result of a fracture of the cervical section of the spine, cervical vertebrae (2) are dorsally displaced (3).

Nerve impulses go from the spinal cord to all the body's organs through the peripheral nerves and regulate their functions. Information from the organs and tissues enters the central nervous system by sensitive nerve fibers. Most of the nerves of the body are composed of sensory, motor and vegetative fibers.

Nerve roots are a short distance to the spinal canal, and then come out of the spinal canal through foramens.

In human body, as in other vertebrates, there is the segmental innervation of the body. Each segment of the spinal cord supplies the specified area of the body.

Skeletopy spinal segments. The roots form 31–33 pairs of spinal nerves. Segment is a portion of the spinal cord, which corresponds to two pairs of roots (two anterior and two posterior). Therefore, 31–33 pairs of spinal nerves in the spinal cord form 31–33 segments: 8 cervical, 12 thoracic, 5 lumbar, 5 sacral and 1–3 coccygeal segment.

Each segment of the spinal cord corresponds to a certain part of the body (Figure 44).

Letters and numbers represent the segments. The letter indicates a region (part) of the spinal cord. The number corresponds to the number of the segment. Cervical segments (segmenta cervicalia) – CI-CVIII; thoracic segments (segmenta thoracica) – ThI-ThXII; lumbar segments (segmenta lumbalia) – LI-LV; sacral segments (segmenta sacralia) – SI-SV; coccygeal segments (segmenta coccygea) – CoI-CoII.

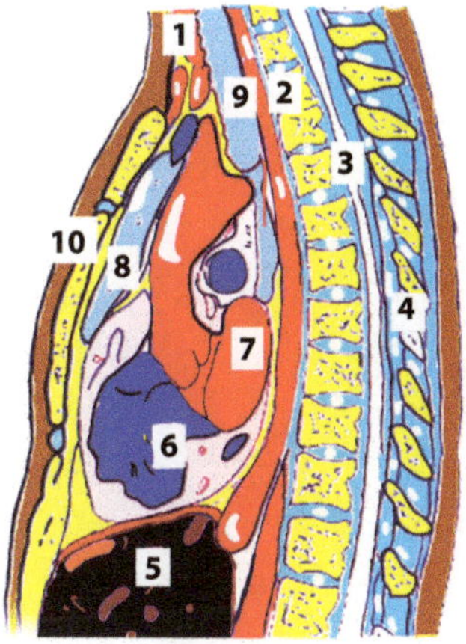

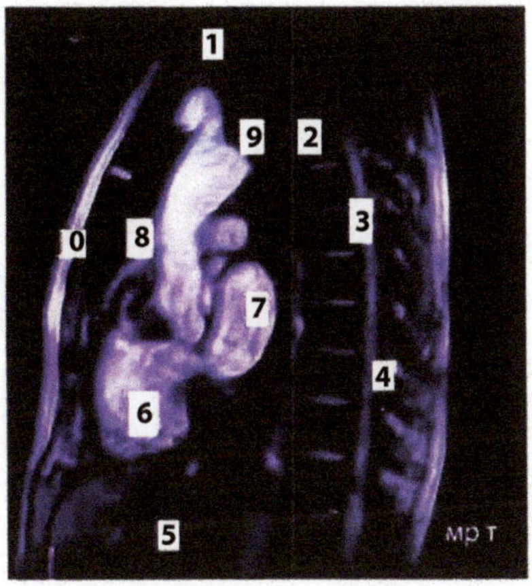

Figure 44 Sagittal section of the chest.
1 – glandula thyroidea; 2 – columna vertebralis; 3 – medulla spinalis; 4 – lig. flavum; 5 – hepar; 6 – ventriculus dexter; 7 – atrium sinister; 8 – pulmo dexter; 9 – trachea; 10 – sternum; 11 – germination tumor site into the vertebral body; 12 – fracture of the upper thoracic vertebra with displacement; 13 – corpus vertebrales; 14 – the displacement of the bodies of the thoracic vertebrae

Spinal cord length is substantially less than the length of the spinal column. Therefore, the sequence number of a segment of the spinal cord and the level of its provisions, starting with the inferior cervical spine, do not correspond to the ordinal number of the vertebra. Segment position relative to the vertebrae can be determined as follows.

The superior cervical segments are arrange at the level appropriate to their ordinal number of the vertebral bodies. The superior thoracic segments lie on one vertebra higher than the corresponding vertebral body.

On average, in the thoracic region the difference between the corresponding segments of the spinal cord have increased by 2 vertebrae in the inferior thoracic – 3.

Lumbar segments of the spinal cord lies in the spinal canal at the level of body X and XI of the thoracic vertebrae, the sacral and coccygeal segments – at the level of XII thoracic and 1 lumbar vertebrae.

Blood supply, venous outflow. Blood supply of the spine is carried out by paired arteries. The cervical vertebral artery departs from the subclavian, goes ahead from of costa-transverse processes of

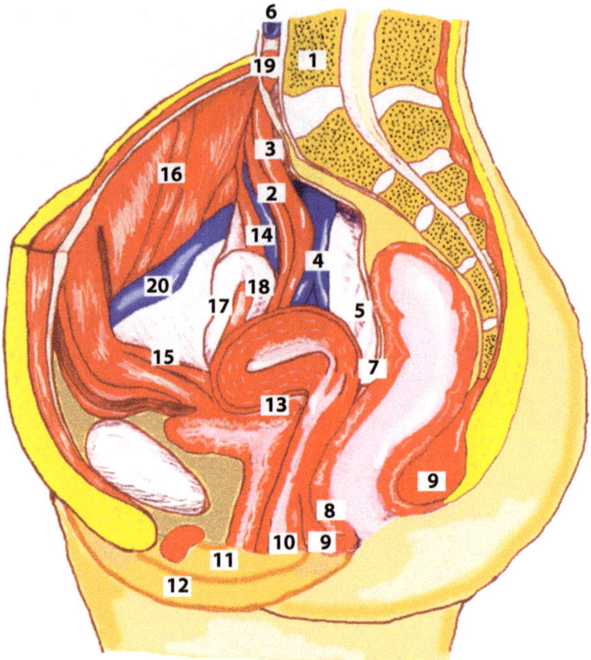

Figure 45 Sagittal section of the female pelvic organs.
1 - cape; 2 – ureter dexter; 3 – a. iliaca interna; 4 – v. iliaca interna; 5 – plica rectouterinalis; 6 – v. comunnis iliaca sinistra; 7 – excavatio rectouterinalis; 8 – m. sphincter ani internus; 9 – m. sphincter ani extemus; 10 – vagina; 11 – labium minus pudendi; 12 – labium majus pudenda; 13 – excavation vesicouterinalis; 14 – vasa ovarii (in supporting ligament); 15 – lig. teres uteri; 16 – a. iliaca externa; 17 – tuba uterine; 18 - ovarium; 19 – a. iliaca comunnis dextra; 20 – v. iliaca externa; 21 – Leiomyosarcoma of uterus; 22 – vesica urinaria; 23 – rectum

C7 vertebra, enters into the canal of the vertebral artery at the level of the C6 vertebra transverse foramens in the channel and goes up (Figure 45).

At the level of the lateral foramens of C2 vertebra vertebral arteries are deflected outwards and go into transverse foramen of the atlas, bent sharply, entwine the back of the atlanto-occipital joint and go into the notch of the vertebral artery on the superior surface of the posterior arch of the atlas. Coming out of it, artery abruptly bent backwards the posterior surface of the atlanto-occipital joint, pierce the posterior atlanto-occipital membrane and go to the notch of a.vertebralis on the

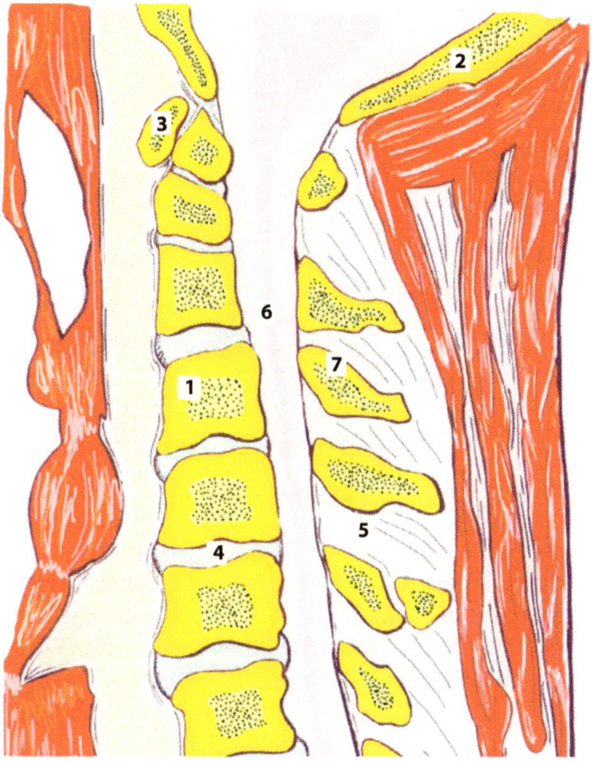

Figure 46 Sagittal section of the cervical processes of the vertebrae.
1 – corpus of axis; 2 – occipital bone; 3 – anterior arch of atlas; 4 – intervertebral disc; 5 – interspinal ligament; 6 – spinal cord; 7 – spinous process

superior surface of the posterior arch of the atlas. Then artery goes through the foramen magnum into the cranial cavity, which is connected to a. basilaris and together with others forms arterial Viliziev circle (Figure 46) – MR-angiography can clearly see that a. carotis interna (13) gives a. communicans posterior (3), which anastomoses with a. cerebri posterior (12). Terminal branches of a. carotis interna (13) are a. cerebri anterior (16) and media (1).

A. cerebri anterior sinistra et dextra approach each other and connect to each other by a. communicans anterior (15). A. cerebri anterior supplies blood to the medial surface of the frontal, parietal and occipital lobes, partially to the hemispheres of the brain (19). The larger a. cerebri media (1) supplies the lateral surface of the same lobe (20)).

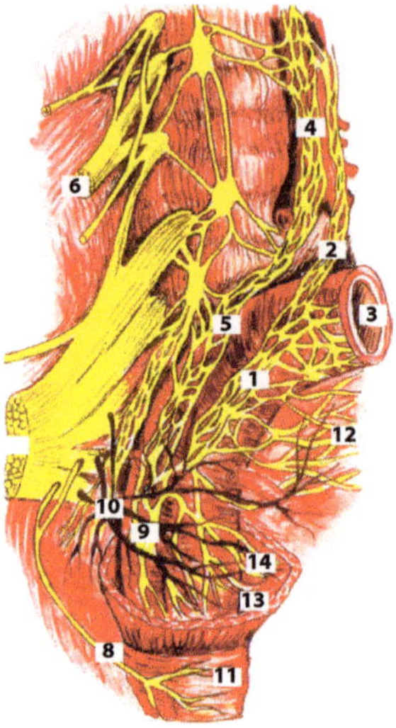

Figure 47 Pelvic plexus.
1 – plexus rectalis medius; 2 – plexus rectalis superior; 3 – rectum; 4 – plexus aorticus; 5 – hypogasricus superior; 6 – plexus lumbalis; 7 – plexus sacralis; 8 – n. pudendus; 9 – hypogasricus inferior; 10 – nn. pelvis interni; 11 – m. sphincter ani externus; 12 – plexus vesicalis; 13 – m. levator ani; 14 – plexus proslaticus

Anterior spinal artery supplies blood to approximately 80% of the diameter of the spinal cord: anterior and lateral cords of white substance, front and lateral horn of the spinal cord, the base of the dorsal horn, brain tissue around a central channel partially – postaerior cords of white substance. Posterior spinal arteries provide blood to the posterior horns of the spinal cord, most of the posterior cords and dorsal portions of the side ropes. Gaulle fasciculus is provided by blood from the basin of the left and right posterior spinal arteries. Burdach fasciculus is supplied with blood from an artery only its part. The blood supply is worse to sections, which are located in critical areas. This area between the anterior and posterior basin spinal arteries: the base posterior horns, brain substance circumference of the central channel, including the posterior commissure and the Clark nucleus. Thus, blood supply to the spinal cord is segmental. There are additional radicular arterys: the spinal branch of the fourth intercostal arteries, spinal branch 11–12 intercostal artery (Adamkevich artery) and inferior external radicular artery (Deprozh-Getteron artery). Last departs

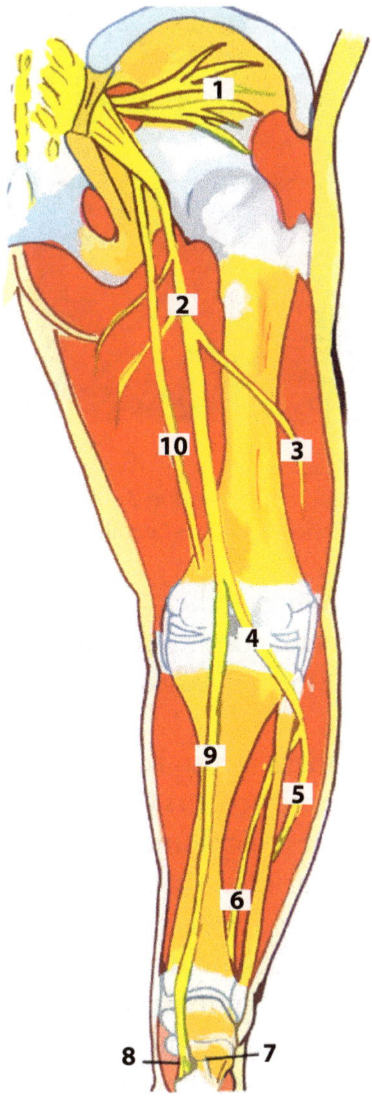

Figure 48 The sciatic nerve and its branches.
1 – n. gluteus superior; 2 – n. ischiadicus; 3 – r. muscularis n. ischiadici; 4 – n. peroneus communis; 5 – n. peroneus superficialis; 6 – n. peroneus profundus; 7 – n. plantaris lateralis; 8 – n. plantaris medialis; 9 – n. tibialis; 10 – n. cutaneus femoris posterior

from the internal iliac artery. Deprozh-Getteron artery with one of the caudal lumbar spinal nerve and its roots reach cone, epicone and spinal cord. Radicular arteries play a leading role in the blood supply to

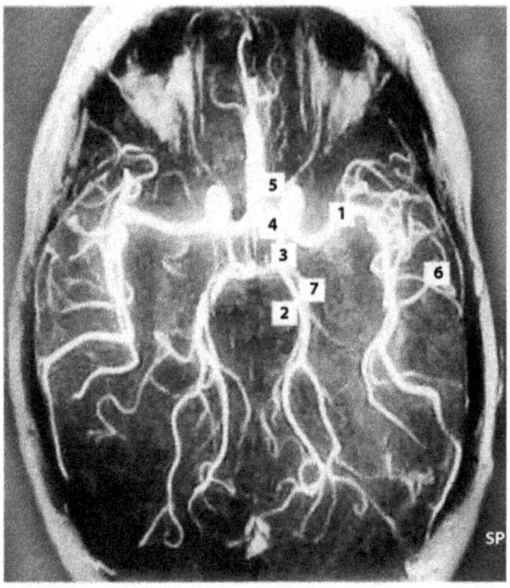

Figure 49 Vessels of the brain. MRI.
1 – a. cerebri media; 2 – a. communicans posterior; 3 – facies medialis; 4 – a. cerebri anterior; 5 – a. communicans anterior; 6 – facies lateralis; 7 – a. cerebri posterior

the spinal cord and its elements. Other spinal branches have secondary importance, but under certain conditions, such as lack of blood flow in one of the main branches, these arteries are involved in compensating circulatory disorders.

Along the spinal cord there is also less reliable blood supply to the areas. They are located on the borders of pools of additional radiculomedullary arteries. Number radiculomedullary arteries and entry level of the spinal cord is very variable, so the location of the critical areas is not the same in different subjects. The most common these areas include the superior thoracic segments 5–7, part of the brain above the lumbar enlargement and the terminal portion of the spinal cord.

The main part of venous blood of the vertebral corpuses is collected in reservoirs. It goes to the back of the vertebral corpuses, leaves it and joins to the anterior internal vertebral plexus. A smaller part of the veins of the vertebral body goes out through the foramens and nutrient flows into the anterior external venous plexus. Venous blood from the vertebral arches is going into external and internal posterior spinal venous plexus. Transverse branches connect the right and left of the anterior internal venous plexus. They form a ring and venous anastomoses with the posterior internal venous plexus. Internal and external venous plexus also anastomose and form umbra and posterior intercostal branches. Lasts fall in azygos vein and hemiazygous vein, which anastomoses but it is linked to the system of superior and inferior venae cavae. Superior 2–5 lumbar veins also flow into the azygos and hemiazygous veins. These veins carry blood in the superior vena cava. The inferior lumbar veins are 2–3 caudal and form short and thick trunk of the iliopsoas. It falls into the common iliac vein. Thus, the venous plexus of the spine are cava-caval anastomosis. In case of insufficiency of blood, flow in the inferior vena cava, pressure of the inferior lumbar vertebral plexus can significantly increase. This leads to varicose veins of the spinal canal, venous congestion, disruption of trophic tissue of the spinal segment and spinal nerve roots of cauda equina and spinal cord, even cone. Vein intervertebral foramens are anastomoses between the internal and external venous plexus. Each intervertebral foramen has 4 vein, one artery and spinal nerve. The blood from the spinal cord enters the radicular veins. They fall into the vertebral venous plexus or directly into the vertebral veins. Arterio-venous anastomoses are between the arterial and venous system. Such arterio-venous shunts are in all tissues and organs. They play an important role in the regulation of blood supply. However, in the spinal cord, they sometimes transform the nature of the vascular malformations. Massive bleeding of arterial blood in the venous canal is insufficient venous return, varicose veins and swelling, dystrophy, degenerative changes in the spinal cord.

Surgical Anatomy of the Malformations of the Spine and Spinal Cord

Among the malformations, the wedge vertebrae and hemivertebrae (complete and incremental); the splitting of the vertebrae to form butterfly vertebrae, vertebral synostosis, arches, articular processes their number, synostosis) are occurred. Such defects aggravate chest deformity; reduce its volume and an excursion (Figure 50). In the cervical region a serious defect is reducing the number of vertebral synostosis them to form a single bone block (Klippel-Feil disease).

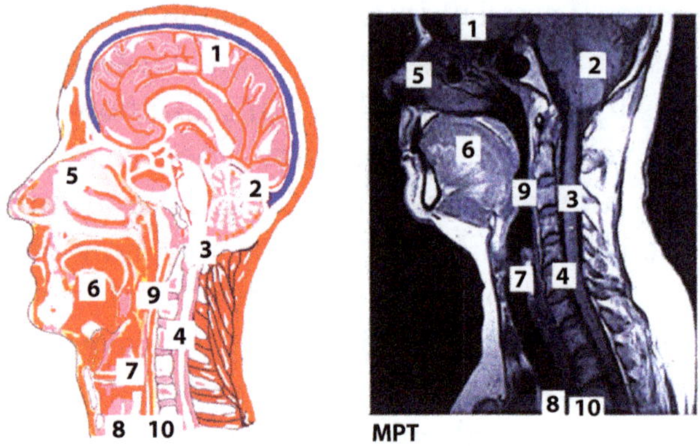

Figure 50 Sagittal section of the head and neck.
1 – hemispheriae cerebri; 2 – cerebellum; 3 – medulla spinalis; 4 – columna vertebralis; 5 – cavum nasale; 6 – lingua; 7 – larynx; 8 – trachea; 9 – pharyngs; 10 – esophagus

Congenital wedge in the thoracolumbar spine can cause severe kyphosis or scoliosis. Splitting arcs of the vertebrae in the lumbosacral refers to variants of the normal ossification. In other parts of the spine vertebrae arc splitting is considered as a sign of pathological development. The lumbosacral splitting of the vertebral corpuses and arches can be combine with malformations of the spinal cord and its roots (myelodysplasia, spinal hernia). It is accompanied by trophic disorders, paresis, contractures and deformities of the lower limbs, disorders of the pelvic organs. The so-called transitional vertebrae are morphologically similar to the vertebrae of another department. It represents the boundary of the vertebrae of the spine. It is accompanied by a corresponding decrease or increase in the number of vertebrae adjacent divisions. In the cervical and thoracic sections, they are not clinically manifested. The lumbosacral transitional vertebrae (sacralization and lumbarization) often contributes to early development of degenerative processes in the spine. Sacralisation of the V lumbar vertebra is more common. Extension and expansion of the transverse process on one or both sides (incomplete form of transitional vertebrae), the deformation of the transverse process usually characterize it. This process coupled with the lateral sacral mass through the cartilage or connective tissue. The bony vertebral usually is not a source of pain. The cartilage and connective connection can develop osteoarthritis, which is accompany by pain. Lumbarization is rare. Diagnosis of malformations of the spine requires an X-ray examination.

Pathotopography of the Spine

Abscess is a formation bounded by an infiltrative capsule, inside of which there is a cavity containing purulent exudate. Abscess does not tend to spread to surrounding tissues.

Pathography (Figure 51). **Retropharyngeal abscess** is formed as a result of suppuration of the lymph nodes and tissue of the pharyngeal space. The pathogens penetrate the lymphatic system from the side of the nasal cavity, nasopharynx, auditory tube and middle ear. Retropharyngeal abscess is classified according to location.

There are following types of pharyngeal abscesses:

- epipharyngeal, located above the palatine curtain;
- mesopharyngeal, localized between the root of the tongue and the edge of the palatine curtain;
- hypopharyngeal, located below the root of the tongue;
- mixed, occupying several anatomical zones.

Abscess is localized in the retropharyngeal space, which is located behind the pharynx. It is limited posterior to the prevertebral, in front of the peripharynx fascia, laterally - pharyngeal-vertebral fascial spurs. At the top, it starts from the base of the skull, at the bottom it passes into the fiber, located behind the esophagus (behind the tissue, the cellular tissue space of the neck), the latter passes into the tissue of the posterior mediastinum. In addition to fiber, the pharyngeal tissue contains single lymph nodes. With the formation of pharyngeal abscesses purulent process can quickly spread along the loose fiber, spatium retroviscerale, in the posterior mediastinum with the development of dangerous posterior mediastinitis. With X-ray examination of the pharynx in the lateral projection of the inflammatory process in the pharyngeal space is characterized by the expansion of its shadow; the pharyngeal abscess is detected as a limited shadow in a certain area.

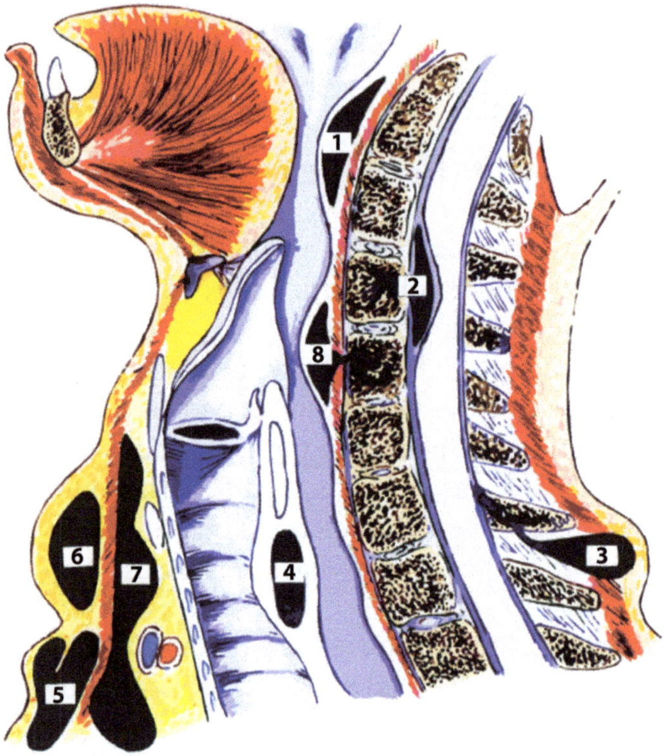

Figure 51 Typical localization of abscesses and phlegmon of the neck.
1 – retropharyngeal abscess; 2 – extradural abscess; 3 – abscess nuchal region; 4 – retrotrahealy abscess; 5 – presternal abscess; 6 – interaponevrosis episternal abscess; 7 – abscesses of previstseral space; 8 – retroesophageal abscess half of body; including pelvis pathology of the heart and blood vessels leads to brain damage.The process begins with a stage of small narrowing, and ends with complete obstruction (occlusion, occlusion) of the vessel

Retrotracheal abscess is an inflammatory process localized between the trachea and the esophagus. An abscess displaces the trachea anteriorly, resulting in a narrowing of its lumen. There is also a compression of the esophagus. Purulent exudate squeezes the recurrent laryngeal nerves, lower thyroid arteries.

Posterior esophageal abscess is most often located in the slit-like retrovisceral space filled with loose fiber and extending from the base of the skull to the posterior mediastinum up to the diaphragm. There is compression of the long muscles of the neck, prevertebral fascia.

The abscess also squeezes the thoracic duct, the right intercostal arteries, the end sections of the semi-nephary and additional semi-neparic veins.

Epidural abscess - accumulation of pus between the dura mater and bone. Epidural abscess often develops in the mid-chest and lower lumbar regions, where the epidural space is best expressed. An offset of the dura mater of the spinal cord is observed. The formation of an abscess leads to compression of the spinal roots, and then the spinal cord.

Pre-tracheal abscess - accumulation of pus between the parietal and visceral sheets of the 4th fascia. With a massive inflammatory process, compression of the main neurovascular bundle of the neck is possible, surrounded by the vagina carotica, which is formed by the parietal sheet of the 4th fascia. An abscess causes the thyroid gland to move forward. It is also possible its involvement in the inflammatory process.

Intervertebral hernia is a pathology in which a protrusion of the jelly-like part of the intervertebral disc occurs beyond its limits.

Pathography (Figure 46). Cervical segment is the most dangerous localization of hernia formation. It is in this area that the centers of the spinal cord are located, which give rise to the nerves going to the respiratory muscles (to the main - the diaphragm, and to the additional - intercostal respiratory muscles). Paralysis of this musculature means stopping the flow of air into the lungs. If help arrives in time, the person can still be saved by connecting to a respirator. In the spinal cord (in its cervical region) there are also other important centers that give commands to the heart, esophagus and hands. The vertebrae of this area are closely in contact with the carotid arteries, jugular veins and their branches, which provide nourishment to the organs of the neck, head, and most importantly - the brain. If a hernia in the cervical region "shoots" between these vertebrae - ischemic stroke may develop.

Carotid stenosis is a disease that binds cardiology and neurology. This is a vivid example of when the pathologyof the heart and blood vessels leads to brain damage.The processbegins with a stage of small

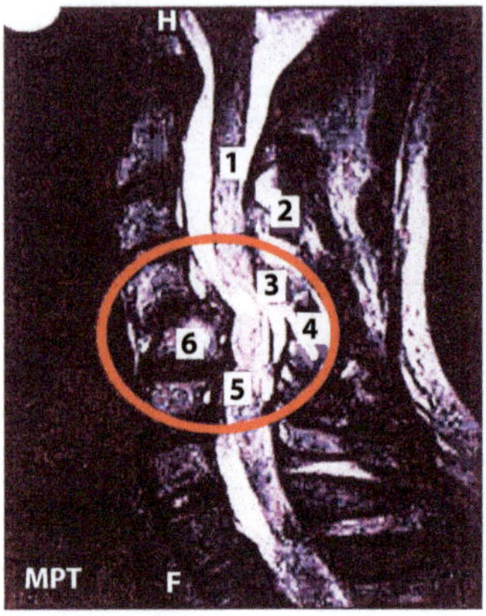

Figure 52 **MRI**. MRI of the cervical spine (hernia).
1 – spinal cord; 2 – spinous processes; 3 – solid membrane of the spinal cord; 4 – interspinal gaps; 5 – herniated intervertebral; 6 – vertebral body

narrowing, and ends with completeobstruction (occlusion, occlusion) of the vessel.

Causes: The carotid artery supplies blood to the cerebral vessels. Internal branch is one of the main components of the structure of the circle of Willis on the basis of the brain. The external artery provides the work of anastomoses in case of circulatory failure; therefore its healthy state is important for the course, severity and prognosis of ischemia. The main blood flow goes along the left and right common trunk, then along the internal carotid artery.

Pathography. Figure 52 shows the pathological tortuosity of the right internal carotid artery (1), the right external carotid artery (2), resulting from stenosis (3) of the right internal carotid artery. These vessels intersect with each other (4), as a result of which the right internal carotid artery shifts to the anterior neck region and partially covers the posterior abdomen of the digestive muscle (5); the right external carotid artery shifts to the lateral processes of the cervical vertebrae, which is not normally observed, a.carotis interna, moves away from the common carotid artery at the level of the upper edge of the thyroid cartilage. It does not enter the cranial cavity through the sleepy canal, but goes to the front of the skull, where it disintegrates into final branches. In cases of absence of the internal carotid artery, the lack of

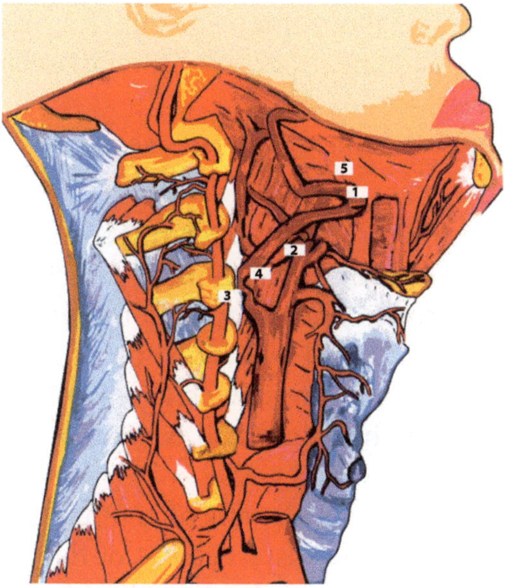

Figure 53 Neck pathography.
1 – a. carotica interna; 2 – a. carotica externa; 3 – stenosis of a. carotica internae; 4 – cross of a. carotica interna et a. carotica externa; 5 – m. digastricus venter posterior

blood supply to the brain is compensated by the much greater development of the corresponding arteries in the opposite hemisphere, as well as by the extraordinary development of the vertebral arteries.

Fracture of the thoracic spine is a pathological condition resulting from a violation of the anatomical integrity of one or more thoracic vertebrae.

Fractures of the thoracic vertebrae resulting from high-energy injuries are often combined with other injuries: rib fractures, fractures of the pelvis, fractures of limb bones, TBI, chest damage, kidney damage, blunt abdominal trauma, etc. There are traumatic and pathological, single and multiple spinal fractures. Traumatic fractures occur with the intense influence of external forces on the unchanged spine. Pathological damage develops in the presence of diseases that reduce the strength of bone tissue, and is formed with minimal external influence or when performing ordinary household actions.

Pathotopography (Figure 53). A spinal fracture has always been considered severe damage to the skeleton. Of the total number of injuries, such injuries are about three percent. They can be combined with damage to the ligaments, roots, muscles, discs and spinal cord, which are located nearby.

Types of fractures of the thoracic vertebrae.

1. Compression fracture. Most often, the victim is the compression of the bodies of the thoracic vertebrae into each other due to strong compression. Compression fractures are usually accompanied by a violation of the integrity of the spinal cord membranes, which greatly aggravates the clinical picture of the injury and is an unfavorable factor for recovery;
2. Rotational fracture. As a result of trauma, a person not only has a fracture and displacement of the thoracic vertebrae, but also damage to the discs, ligaments, as well as joint surfaces and ribs;
3. Distraction of the fracture. The victim has an excessive stretching of the vertebral column, which is accompanied by a violation of the integrity of the body and processes of the vertebrae, as well as a rupture of the back muscles and intervertebral discs.

On sagittal sections of the thorax, pathotopography differentiates the germination of the tumor node in the vertebral body, the fracture of the upper thoracic vertebrae with displacement. Due to a serious compression fracture of the spine in the future, the patient may experience segmental instability, in which pain is concerned, degenerative changes occur.

In elderly people who have suffered a spinal fracture, there may be "old hump" and chronic pain in the spine.

Spinal tumors are a neoplasm that occurs in the cells of the spinal column and grows both inside the spinal canal and outside, to the adjacent organs and tissues. This disease includes a variety of tumors, which differ in the structure of cells, which consist of. The prevalence in different countries of the world is about the same. As such, the cause

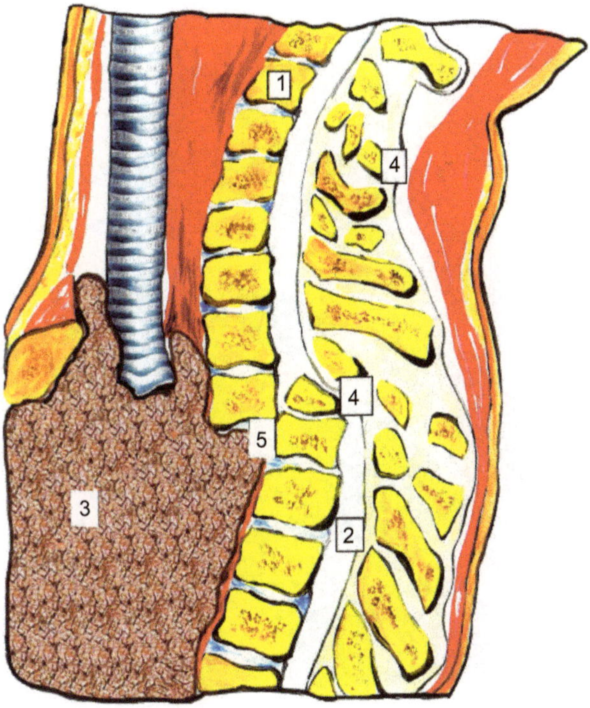

Figure 54 Sagittal section of the chest.
1 – columna vertebralis; 2 – medulla spinalis; 3 – pulmo dexter; 4 – the upper thoracic vertebrae fracture with displacement; 5 – displacement bodies of the thoracic vertebrae

of any spinal tumor is not found. There is an assumption that this is a genetic defect that occurs during the release of cells during fetal development. Such risk factors include exposure to radiation, long Smoking experience (including passive), systematic violation of the diet (incorrect ratio of daily consumption of proteins, fats and carbohydrates), frequent and prolonged exposure to open sunlight or artificial ultraviolet (tanning, etc.), work on chemical production (paints, resins, solvents, etc.).)

Pathotopography (Figure 54). Cancer of the cervical spine can be determined by the appearance of non-specific signs of cancer and neurological disorders.

1. Local manifestations – moderate pain, first concentrated only in the place of tumor growth, then extends to other

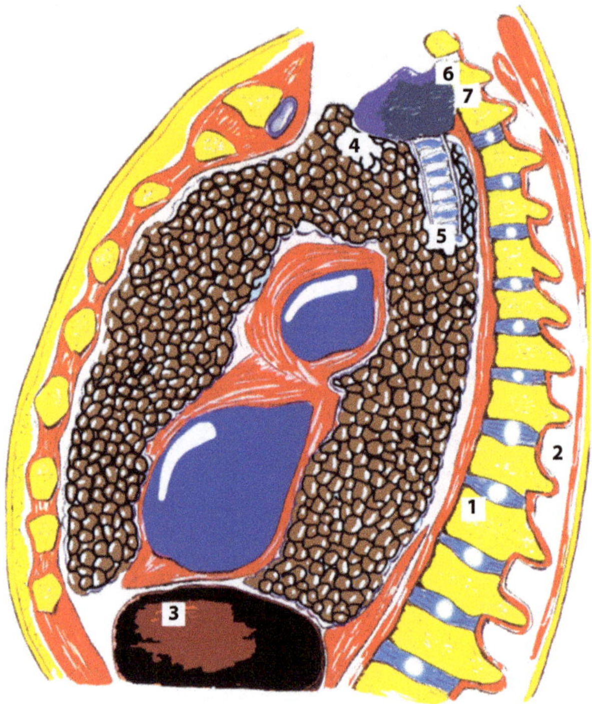

Figure 55 Sagittal section of the chest.
1 – columna vertebralis; 2 – lig. flavum; 3 – hepar; 4 – pulmo dexter; 5 – trachea; 6 – germination tumor site into the vertebral body; 7 – corpus vertebralis

parts of the spine. The nature of the pain at the initial stage of the malignant process is moderate, but then other symptoms join, and only narcotic analgesics can get rid of the pain.

2. Paresis of a limited area of the skin depends on the nerve that compresses the cancer. Most often, paresis occurs in the upper part of the hand, but the loss of sensitivity of other parts of the body is not excluded. In this case, the patient feels numbness, tingling or a complete lack of response to tactile, temperature stimuli.
3. Paralysis – partial or complete absence of movement. In the process of cancer development there is a smooth transition from partial to complete loss of limb movement.

It is important to note that the higher the localized cancer, the more tissue and spine affected. The development of the tumor in the cervical spine leads to more pronounced symptoms, the patient feels terrible pain throughout the spine, and then the function of individual organs is disturbed. Cancer of the cervical spine is accompanied by a violation of vestibular function. There are problems with coordination and management of the body, muscle spasms and fasciculations. On the part of the renal system, there may be a violation such as urinary incontinence or anuria. The patient has increased blood pressure, disturbed heart rate, there is a noise in the ears, darkening before the eyes. In the morning, the patient can observe red circles under the eyes, the flickering of flies before the eyes.

Part 3: The Limbs

Topographic Anatomy of the Upper Limb

Human upper limb is divided into four sections: shoulder girdle, shoulder, forearm and wrist.

Supra Brachium – Shoulder Girdle

There are four areas of the shoulder girdle: the axillary, deltoid, subclavian and scapular.

Axillary fossa (fossa axillaris) refers to the axillary region (Figure 56). In front it is limited by pectoralis major muscle, m. pectoralis major, behind – the broad back muscle (m. latissimus dorsi), inside – the serratus anterior muscle (m. serratus anterior) and outside – two muscles of the upper part of the shoulder: short head of the biceps (caput brevem bicipitis brachii), and coracobrachialis muscle (m. coracobrachialis) fused with it.

The neurovascular bundle passes in the depth of the axillary cavity, cavum axillare.

Plexus brachialis (Figure 56) – the brachial plexus located here consists of three bundles – external, internal and posterior.

External bundle (fasciculus lateralis) gives the external root, radix lateralis, for the formation of the median nerve, n. medianus, below the bundle enters the musculocutaneous nerve (n. musculocutaneus) which innervate motor branches of all three flexors which located in

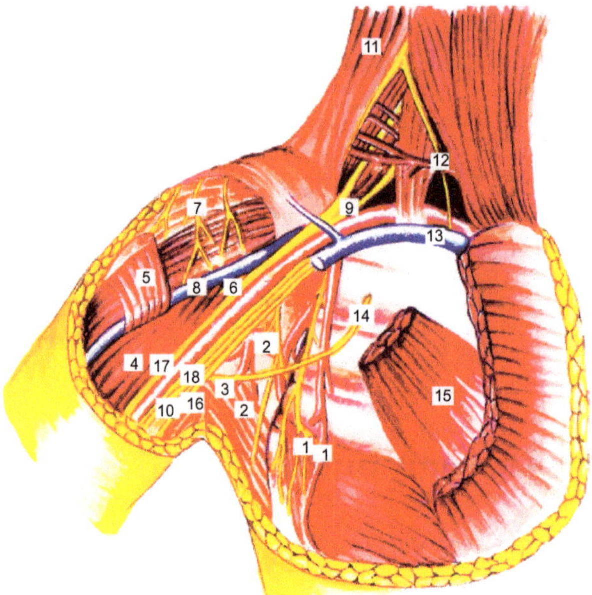

Figure 56 The brachial plexus.
1 – a. thoracica externa; n. thoracicus longus; 2 – m. latissimus dorsi et a. subscapularis; 3 – a. circumflexa humeri posterior; 4 – m. biceps brachii; 5 – m. pectoralis major; 6 – n. musculocutaneus; 8 – n. cutaneus humeri externus; 9 – plexus brachialis; 10 – n. cutaneus antebrachii intemus; 11 – m. trapezoidea; 12 – n. phrenicus et truncus thyrohyoideus; 13 – v. subclavia; 14 – n. intercostobrachialis; 15 – m. pectoralis minor; 16 – n. cutaneus brachii medialis; 17 – n. medianus; 18 – n. ulnaris

the anterior shoulder and the skin of the external part of the front forearm.

Internal bundle (fasciculus medialis) gives the internal root, radix medialis, for the formation of the median nerve and ulnar nerve below (n. ulnaris).

Posterior bundle (fasciculus posterior) forms two nerve:

- the radial nerve (n. radialis) innervates motor branches all extensors shoulder and forearm, and by sensitive branches it innervates skin of the posterior area of all limb;
- the axillary nerve (n. axillaris) innervates the deltoid muscle and skin of the area.

Thus, there are four main nerves in the axillary region of the upper limb:

1. n. musculocutaneus – musculocutaneous nerve, it innervate all three flexors of shoulder;
2. n. medianus – the median nerve, it innervate the lateral flexors of the forearm and the skin of three and a half fingers with radiation side;
3. n. ulnaris – ulnar nerve, it innervate the medial flexor lying on the forearm and the skin of the one and half fingers with the ulnar side;
4. n. radialis – radial nerve, it innervate system of flexor muscles of limbs and skin two and half fingers on the back side, starting from the first finger.

Damage or inflammation of the nerves causes typical symptoms:

1. the radial nerve – the symptom of dangling hand, as a result of lesion of the extensor;
2. the median nerve – the symptom of "monkey paw" due to atrophy of the muscles of the thumb elevation;
3. the ulnar nerve – the symptom of "clawfinger hand", coming as a result of atrophy of the intercostals muscles (the patient can not push his fingers to the side);
4. the musculocutaneous nerve – a patient can not bend the forearm.

The topography of the axillary artery (a. axillaris) (Figures 56 and 57). The axillary artery begins at the outer edge of first rib and extends to the outer edge of the pectoralis major muscle. Coming here, it turns into the brachial artery. Its projection on the front edge of the hairy part of axilla. It is surrounded by three nerves: the median nerve is adjacent to it on the outside (and even lateral musculocutaneous nerve). The ulnar nerve is adjacent inside, and the radial nerve behind. It is important to know when the operation of the vessel ligation or suturing nerve ends crossed Conditionally artery is divided into three

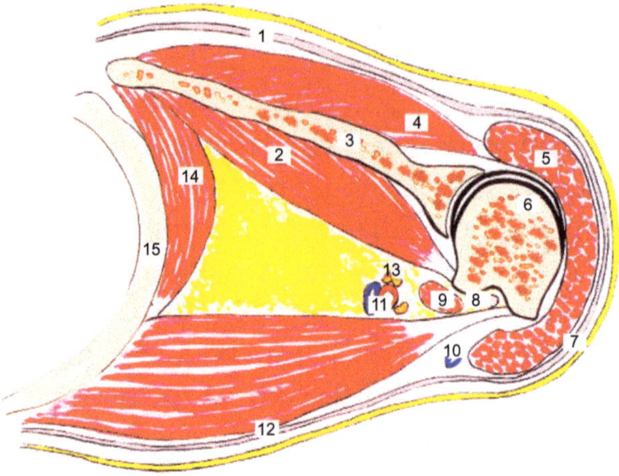

Figure 57 The axillary area.
1 – fascia infraspinata; 2 – m. subscapularis; 3 – scapula; 4 – m. infraspinatus; 5 – m. deltoideus; 6 – caput humeri; 7 – fascia deltoidea; 8 – tendo m. biceps brachii; 9 – m. coracobrachialis; 10 – v. cephalica; 11 – a. et v. axillaris; 12 – m. pectoralis major et fascia thoracica; 13 – plexus brachialis; 14 – cellulosum fossae axillaris; 15 – m. serratus anterior; 16 – costa secunda (II)

parts: before leaving under the pectoralis minor, behind it and after getting out from under her.

Axillary artery gives six branches. In the first segment of two vessels: upper thoracic artery (a. thoracalis suprema), the artery of the chest and the shoulder bone (a. thoracoacromialis). They supply blood to the major and minor pectoral muscles. In the second segment extends one vessel – a. thoracalis lateralis – the blood supply to the serratus anterior muscle. In the third segment – three vessels: anterior artery surrounding the shoulder (a. circumflexa humeri anterior), posterior artery surrounding the shoulder (a. circumflexa humeri posterior) and subscapularis artery (a. subscapulars). They all supply blood to the muscles of the shoulder, deltoid and subclavian areas.

In the armpit (Figure 58) are arranged axillary lymph nodes (nodi axillares). They lymph flows from the entire upper limb (Figure 55). When festering within the fingers, hands or forearms axillary lymph nodes swell, and sometimes purulent melted, causing a dangerous deep axillary phlegmon. Sometimes the inflamed and axillary sweat gland, which is called hidradenitis.

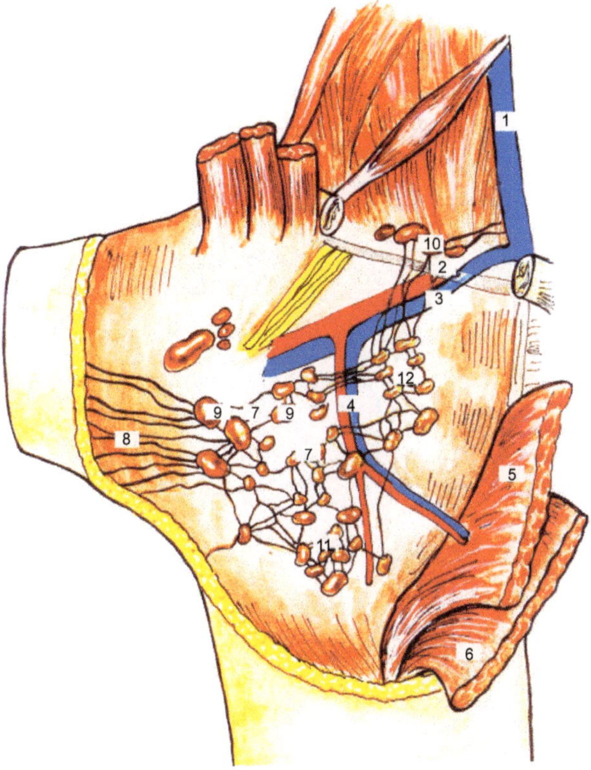

Figure 58 Lymphatic vessels of axilla.
1 – v. jugularis interna dextra; 2 – truncus subclavius; 3 – v. subclavia; 4 – a. et v. thoracica lateralis; 5 – m. pecioralis minor; 6. – m. pecioralis major; 7 – plexus lymphaiicus axillaris; 8 – lymph vessels of the upper limb; 9 – nodi axillares; 10 – nodi supraclaviculares; 11 – nodi pectorales; 12 – nodi infraclaviculares

Regio deltoidea (Figure 60) – the deltoid region is bounded by the limits of the deltoid muscle. It starts from the scapular spine and acromion process and attaches to the humerus to the deltoid tuberosity. The muscles of the arm is withdrawn to a horizontal position. N. axillaris innervate it.

Regio infraclavicularis (Figure 60) – subclavian area is located below the collarbone. Here is the neurovascular bundle in the following sequence: surface and is located just below the axillary vein; above it and deeper – the axillary artery; higher and deeper is all brachial plexus.

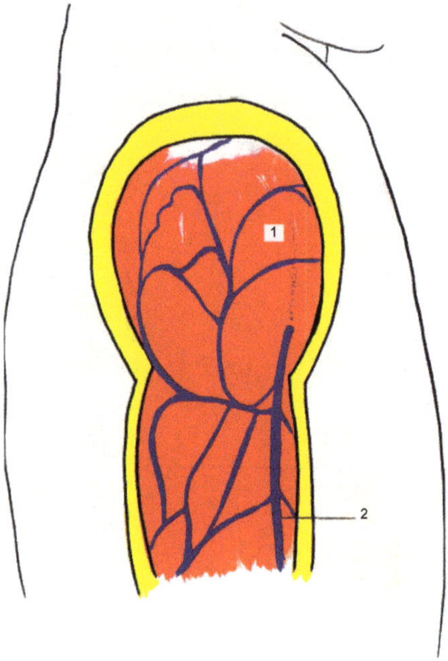

Figure 59 Deltoid region.
1 – v. deltoideus; 2 – v. cephalica

Regio scapularis (Figure 61). Scapular region is limited outside of scapula. Here are located: at the top – the supraspinatus muscle (m. supraspinatus), infraspinatus muscle below (m. infraspinatus), lower – teres minor muscle (m. teres minor), and lower than all – teres major muscle (m. teres major). In front of the blades lies the subscapularis muscle (m. subscapularis). Branches of the brachial plexus and axillary artery supplying the branches innervate all these muscles.

Shoulder

On the shoulder there are two muscular covers. In front flexors lie: the biceps (m. biceps brachii), shoulder muscle (m. brachialis) and the coracobrachialis muscle (m. coracobrachialis). These muscles flex the shoulder and upper arm (Figure 62).

In the rear is muscular bed triceps (m. triceps brachii), it extends the shoulder and upper arm (Figure 63 – MR and CT scan shows

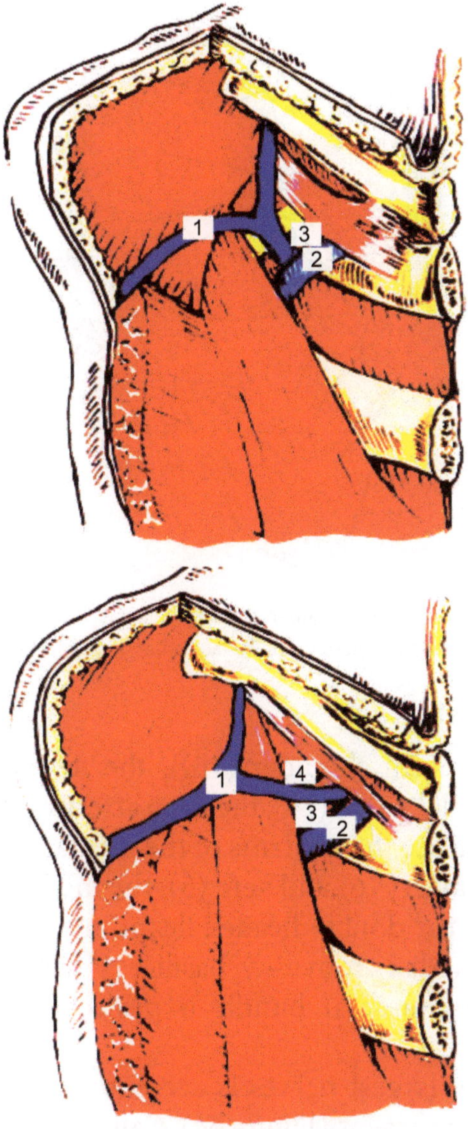

Figure 60 Different forms of trigonum clavipectorale (by D.D. Kupatadze).
a – triangular, high confluence v. cephalica, b – trapezoid, low confluence v. cephalica.
1 – v.cephalica; 2 – v. subclavia; 3 – a. subclavia; 4 – pl. brachialis

all anatomical structures identified in the figure caption, also shows the contents of the shoulder-muscle channel (20) (limited front – humerus (17), posterior and lateral – triceps muscle (18), medially

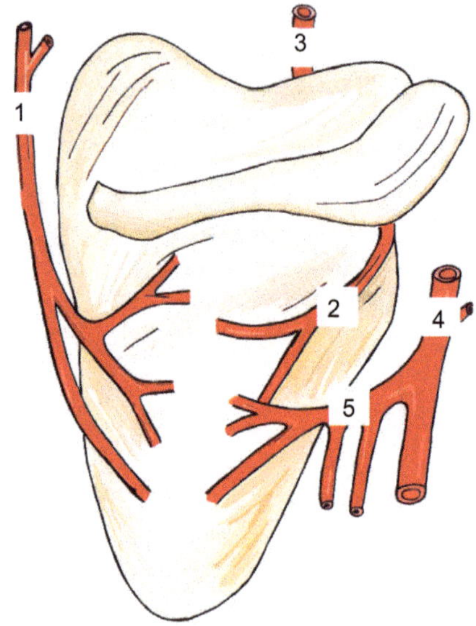

Figure 61 Blood circulation of the scapular region.
1 – a. suprascapularis; 2 – a. transversa colli; 3 – a. axillaris; 4 – a. thoracodorsalis; 5 – a. circumflexa scapulae

– intermuscular septum shoulder (16)), the neurovascular bundle, consisting of the radial nerve (14), the deep artery of the shoulder (13). In addition, you can see the contents of the medial furrow (19), biceps (11) – the brachial artery (6) and vein (5), the median nerve (4)).

On either side of the double-headed flexor in the grooves lie saphenous veins: in the medial groove – v. basilica, in the lateral – v. cephalica. The main neurovascular bundle is projected along the medial groove.

Each of the three largest nerves is accompanied by arterial vessel and by two same-name veins (Figure 64). Canals of the anterior region of the shoulder. Canalis humeromuscularis is located between m. triceps brachii and the spiral groove of the humerus and connects the armpit with the ulnar area. Content – a. profunda brachii et n. radialis (Figure 65).

Shoulder-humeroradial channel located between the brachial and brachioradialis muscles, the contents are n.radialis and radial collateral oblique vessels. For example, the median nerve is accompanied by

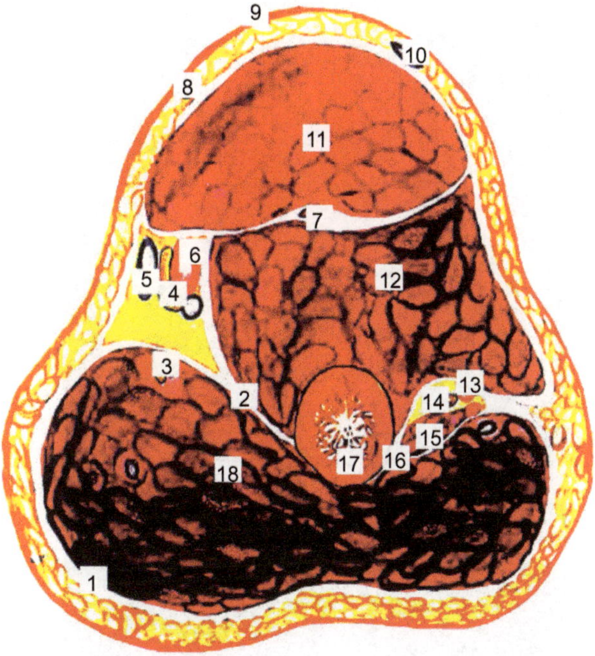

Figure 62 Crosscut of shoulder in the lower third.
1 – fascia brachii; 2 – septum intermusculare internum; 3 – n. ulnaris et a. ulnaris; 4 – n. medianus; 5 – v. brachialis; 6 – a. brachialis; 7 – n. musculocutaneus; 8 – panniculus adiposus; 9 – derma; 10 – v. cefalica; 11 – m. biceps brachii; 12 – m. brachialis; 13 – a. profunda brachii; 14 – n. radialis; 15 – m. brachiradialis; 16 – septum intermusculare externum; 17 – humerus; 18 – m. triceps brachii m. subscapularis, below – m. teres major, laterally is a surgical neck of the humerus. Content on the posterior septum of the axilla – a. et v. circumflexae humeri, n. axillaris

brachial artery (a. brachialis), radial nerve goes along with deep arteries of the shoulder, a. profunda brachii, with the ulnar nerve passes a thin upper ulnar artery, a. collateralis ulnaris superior (Figure 62).

Borders of foramen trilaterum: laterally – a long head of m. triceps brachii, m. teresminor and m. subscapularis from above, teres major muscle below, the contents – a. et v. circumflexae scapulae.

Borders of foramen quadrilaterum medially – are a long head of m. triceps brachii, from above – m. teres minor and

Superficial veins of the upper limbs are shown in Figures 63–65.

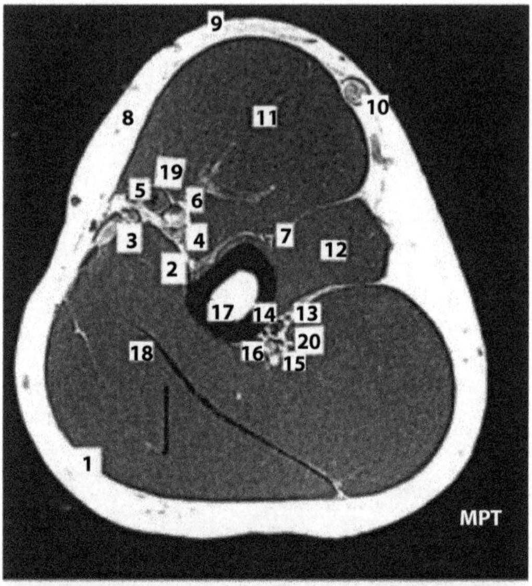

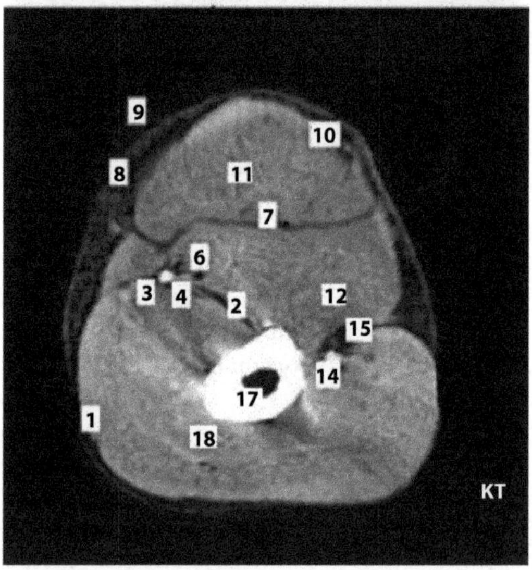

Figure 63 Crosscut of shoulder in the lower third (MRI, CT).
1 – fascia brachii; 2 – septum intermusculare internum; 3 – n. ulnaris et a. ulnaris; 4 – n. medianus; 5 – v. brachialis; 6 – a. brachialis; 7 – n. musculocutaneus; 8 – panniculus adiposus; 9 – derma; 10 – v. cefalica; 11 – m. biceps brachii; 12 – m. brachialis; 13 – a. profunda brachii; 14 – n. radialis; 15 – m. brachiradialis; 16 – septum intermusculare externum; 17 – humerus; 18 – m. triceps brachii

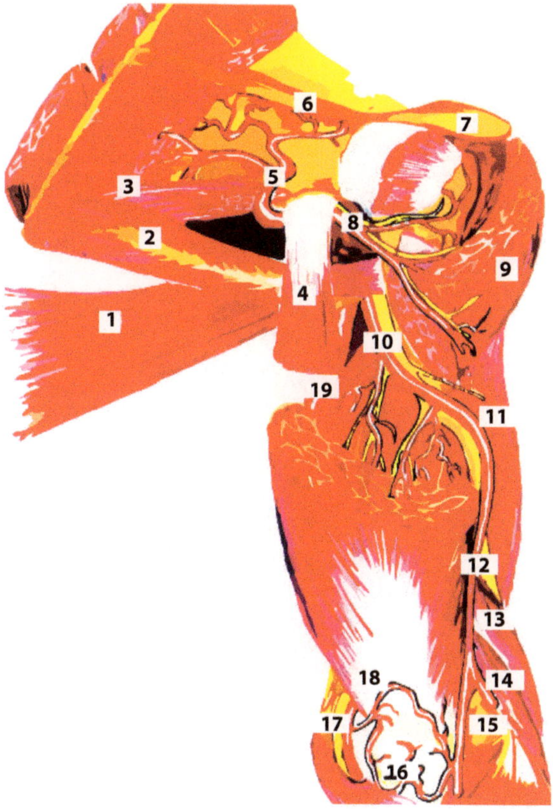

Figure 64 The posterior brachium area.
1 – m. latissimus dorsi; 2 – m. teres major; 3 – mm. infraspinatus et teres; 4 – caput longum m. triceps brachii; 5 – a. circumllexa scapulae; 6 – spina scapulae; 7 – processus acromialis scapulae; 8 – n. axillaris et a. circumllexa humeri posterior; 9 – m. deltoideus; 10 – n. radialis et a. profunda brachii; 11 – m. biceps brachii; 12 – n. radialis et a. collateralis radialis; 13 – m. brachioradialis; 14 – m. extensor carpi radialis longus; 15 – condylus humeri externus; 16 – oleocranon ulnae; 17 – condyius humeri internus et n. ulnaris; 18 – m. triceps humeri et a. collateralis ulnaris superior; 19 – a. brachialis

Forearm

The structure of the forearm and hand is extremely difficult, due to their function: they participate in the work activity. There are a lot of muscles in fascial covers, blood vessels and nerves. To facilitate memorization of the main muscles and their functions can use the following functional diagram dividing them into groups: 6-3-6-4. The first group

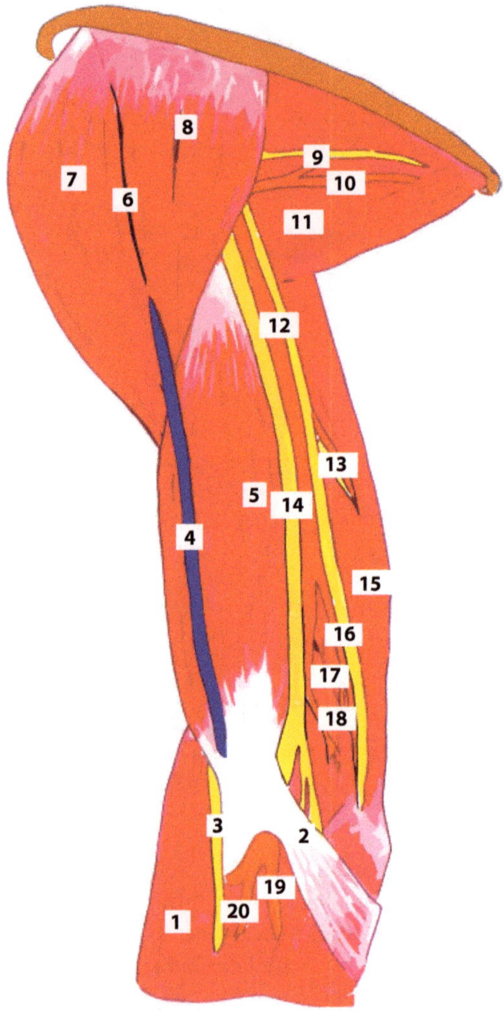

Figure 65 The topography of the anterior brachium area.
1 – m. brachioradialis; 2 – tendo accessorius m. biceps brahii; 3 – n radialis; 4 – v. cephalica; 5 – m. biceps brahii; 6 – sulcus 7 – m. deltoideus; 8 – m. pectoralis major; 9 – a. et v. subscapularis; 10 – a. circumflexa scapulae; 11 – m. teres minor et m. latissimus dorsi; 12 – a. brachialis; 13 – n. radialis et a. profunda brachii; 14 – n. medianus; 15 – capurt inferior m. triceps brachii; 16 – n. ulnaris et a. collateralis ulnaris superior; 17 – septum intermusculare internum; 18 – m. brachialis; 19 – a. ulnaris; 20 – a. radialis

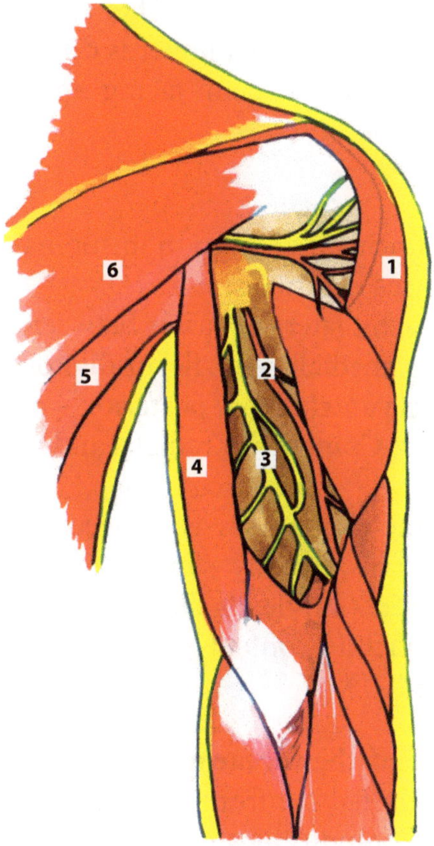

Figure 66 The topography of the radial nerve and the deep artery of the upper arm. Back view.
1 – m. deltideus; 2 – a. profunda brachii; 3 – n. radialis; 4 – m.triceps brachii; 5 – m. triceps major; 6 – m. infraspinatus

includes six muscles that operate only on the wrist, but not acting on his fingers. The second group includes three muscles, each of which is subdivided into four tendons. The third component of the six muscles run to separate fingers.

The last, the fourth group includes the so-called pronators and supinators. Consequently, they all are located on the forearm muscles.

Functionally, the muscles are divided into flexors, extensors, pronators and supinators. Flexors lie on the palm side of the forearm, extensors – on the backside. Pronators and supinators to rotate the hand around the entire axis of the forearm. This pronator wrist medially rotated (palm facing down, the first finger is directed medially); supinators rotate the arm outward (palm facing up, first finger pointing laterally) (Figures 67–69).

Canals of forearm:

1. canalis supinatorius runs deep in m.supinator. Content – ramus profundus n. radialis.
2. canalis antebrachii lateralis passes between: laterally – m. brachioradialis; medially – m. flexor digitorum superficialis; front – fascia antebrachii; posterior – m. flexor digitorum profundus. Content – v.radialis and r.superficialis n. radialis.
3. canalis antebrachii medialis is between: medially – m. flexor carpi ulnaris; lateraly – m. flexor digitorum superficialis; posterior – m. flexor digitorum profundus; front – fascia antebrachii. Content – n. ulnaris and v. ulnaris (Figures 70 and 71).

The cellullar space of the forearm is relatively closed. The space of Pirogov Paron is limited by: front – m. flexor pollicis longus and m. flexor digitorum profundus; posterior – membrane interossea and m. pronator quadratus; lateral – anterior radial intermuscular septum, medially – own fascia of the forearm; at the top of the attachment to the interosseous membrane m. flexor pollicis longus and m. flexor digitorum profundus. In children this space is represented by a narrow slit.

Vessels accompanied by nerves lie in special canals:

1. Canalis antebrachii medialis – internal canal of the forearm is located on the palmar side between the medial and middle third of the forearm. It is limited to the ulnar side – m. flexor carpi ulnaris, to the radial side – m. flexor digitorum superfclalls. The bottom of the canal is m. flexor digitorum profundus. Here the ulnar artery and ulnar nerve lie.

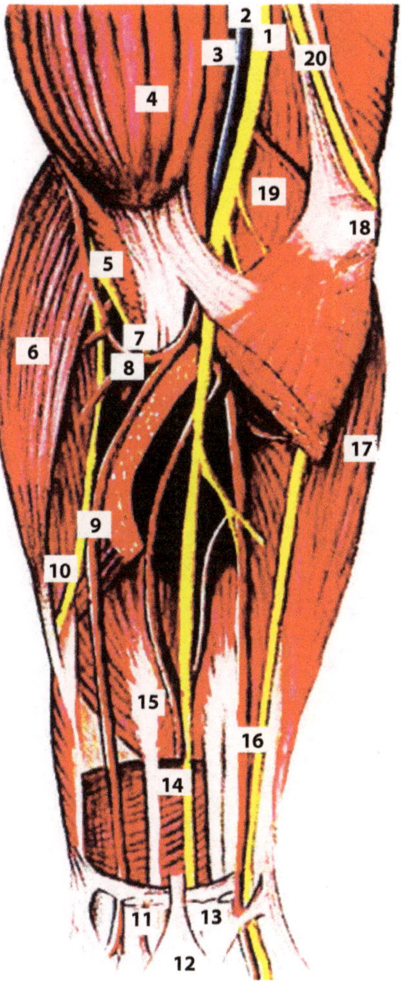

Figure 67 The front surface of the forearm.
1 – n. medianus; 2 – v. brachialis; 3 – a. brachialis; 4 – m. biceps brachii; 5 – n. radialis; 6 – m. brachioradialis; 7 – r. profundus n. radialis; 8 – a. radialis recurrens; 9 – a. radialis et r. superficialis n. radialis; 10 – m. extensor carpi radialis longus; 11 – tendo m.flexorcarpi radialis; 12 – aponeurosis palmaris; 13 – tendo m. flexor digitorum .superficialis; 14 – m. pronator quadratus; 15 – m. flexor pollicis longus; 16 – a. et n. ulnaris; 17 – m. flexor carpi ulnaris; 18 – condylus humeri interims; 19 – m. brachialis; 20 – n. ulnaris on the brachium

2. Canalis antebrachii lateralis – outer canal of the forearm is located between the middle and outer third of the forearm. It is limited to the ulnar side m. flexor carpi radialis

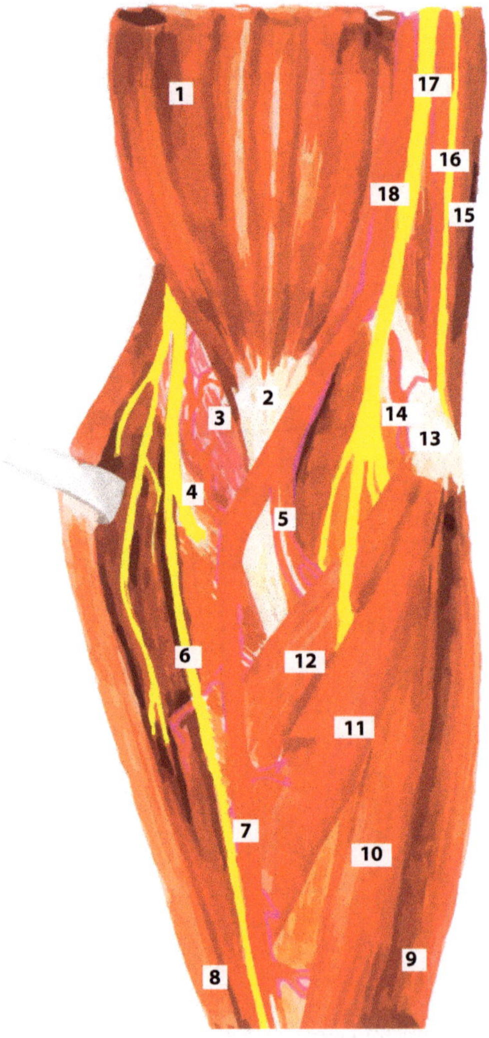

Figure 68 Anterior ulnar area.
1 – m. biceps brachii; 2 – tendo m. biceps brachii; 3 – m. brachialis; 4 – r. profundus n. radialis et a. radialis recurrens; 5 – a.ulnaris; 6 – r. superficialis n. radialis; 7 – a. radialis; 8 – m. brachiradialis; 9 – m. flexor carpi ulnaris; 10 – m. flexor digitorum superficislis; 11 – m. flexor carpi radialis; 12 – m.pronator teres; 13 – condylus humeri interims; 14 – a. collateralis ulnaris inferior; 15 – m.triceps brachii; 16 – n.ulnaris et a. collateralis ulnaris superior; 17 – n. medianus; 18 – a.brachialis

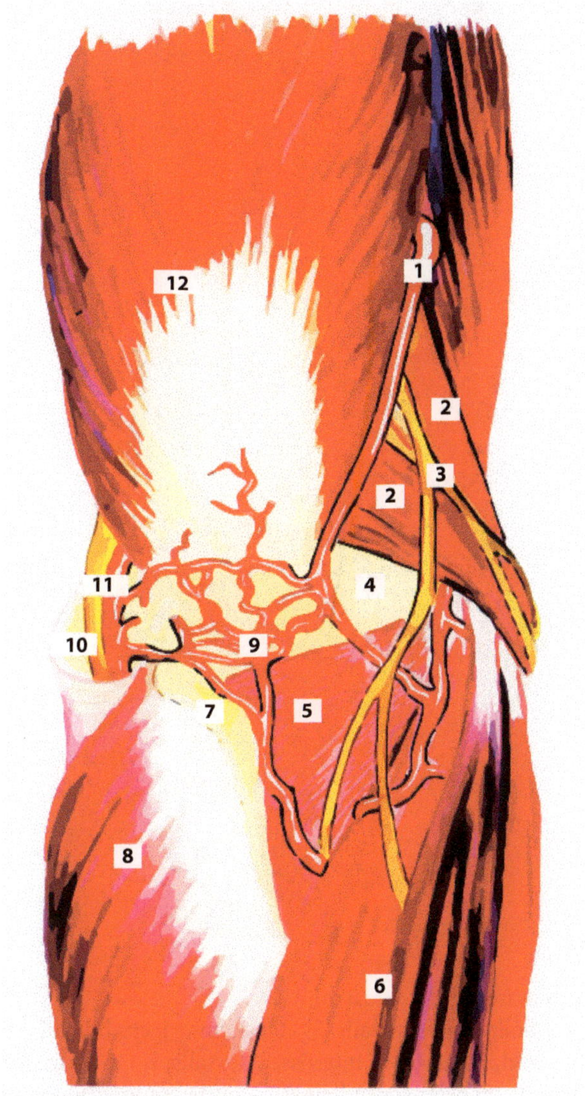

Figure 69 The posterior ulnar area.
1 – a. collateralis radialis; 2 – m. brachioradialis et m. extensor carpi radialis longus; 3 – n. cutaneus antebrachii dorsales; 4 – condylus humeri externus; 5 – m.ulnaris; 6 – m. extensor digitorum communis; 7 – ulna; 8 – m. flexor carpi ulnaris; 9 – rete articulare cubiti; 10 – condylus humeri internus; 11 – n. ulnaris et a. collateralis ulnaris superior; 12 – m. triceps brachii

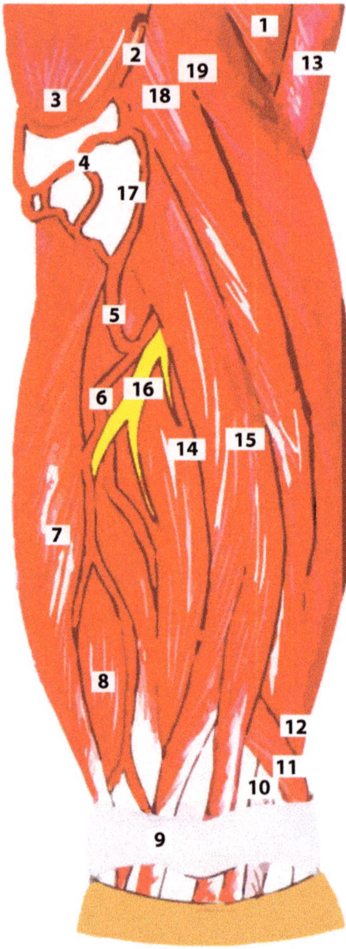

Figure 70 The posterior area of the forearm.
1 – m. brachialis; 2 – a. collateral is radialis; 3 – tendo m. triceps brachii; 4 – rete a. cubiti; 5 – r.interossea recuirens dorsalis; 6 – a. interossea dorsalis; 7 – m. extensor carpi ulnaris; 8 – m. extensor indicis; 9 – m.extensor indicis; 10 – retinaculum extensorum manus 11 – tendo m.extensor carpi radialis longus et brevis; 12 – m. extensor pollicis brevis; 13 – m. abductor policis longus; 14 – m. biceps brachii; 15 – m. supinator; 16 – m. extensor digitorum communis; 17 – r. superficialis n. radialis; 18 – condylus humeri externus; 19 – extensor carpi radialis longus et brevis; 20 – m. braehiradialis the ulnar side – m. flexor carpi ulnaris, to the radial side – m. flexor digitorum superfclalls. The bottom of the canal is m. flexor digitorum profundus. Here the ulnar artery and ulnar nerve lie

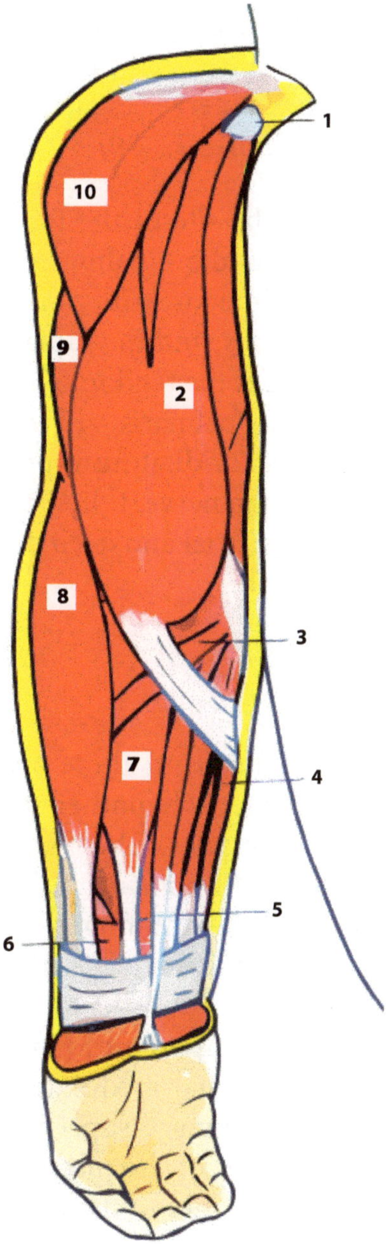

Figure 71 Muscles of the upper limb. Front view.
1 – processus coracoideus; 2 – m. biceps brachii; 3 – m. pronator teres; 4 – m. flexor carpi ulnaris; 5 – m. palmaris longus; 6 – m. pronator quadratus; 7 – m. flcxor carpi radialis; 8 – m. brachioradialis; 9 – m. triceps brachii; 10 – m. deltoideus

and to the radial side m. brachioradialis. Here are located a. radialis and ramus superficialis nervi radialis – superficial branch of the radial nerve (Figures 72–75).Superficial veins are shown in(Figures 77–79)

There are three canals at the site of the forearm transition to the wrist (Figure 80): a larger, middle – palmar canal, canalis palmaris, and two side – ulnar canal of the forearm, canalis carpi ulnaris, where the ulnar vessels and nerve pass, and the radial canal of the forearm, canalis carpi radialis, where lies the tendon of same name muscle.

In the palmar canal, canalis palmaris, are three muscles: m. flexor digitorum superficialis, m. flexor digitorum profundus and m. flexor polllicis longus, and the median nerve. It lies strictly in the midline of the forearm between the superficial and deep digital flexors.

Hand (Figure 81)

On the palmar surface of the hand there are two elevations: the thumb – thenar and fifth finger – hypothenar. Each has four small, but practically important muscles. The muscle that removes the thumb, leading it, flexing and opposing it, lies in the elevation of the thumb; at an elevation of the fifth finger – the same muscles, but there is no adductor muscle, which replaces the short palmar subcutaneous muscle.

For practical purposes, it is important to know the location of the superficial and deep arterial palmar arches. (Figures 82 and 83). The superficial palmar arch, arcus palmaris superficialis, is projected when the thumb is moved to the level of the distal elevation; deep palmar arch, arcus palmaris profundus, lies in the middle of the base of this elevation.

Innervation of fingers and hand is carried out in this way: on the palm side of the three and a half fingers (1, 2, 3 and a half of the fourth) are innervated by the median nerve; the other part – is due to the ulnar deviation (Figures 80 and 81).

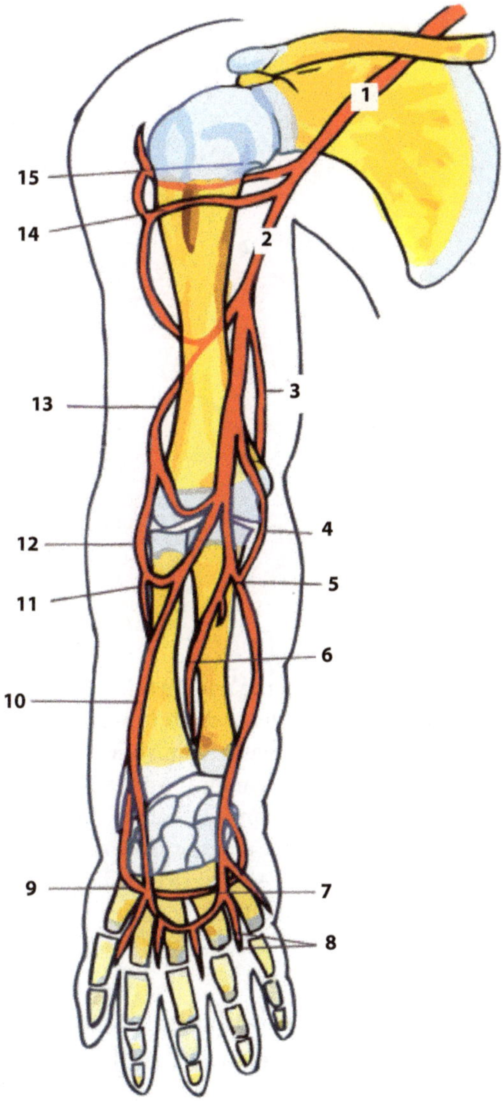

Figure 72 Arteries of the upper limb.
1 – a. axilaris; 2 – a. brachialis; 3 – a. collateralis ulnaris inferior; 5 – a. recurrens ulnaris; 6 – a. ulnaris; 7 – a. interossea anterior; 8 – arcus palmaris superficialis; 9 – aa. digitales palmares propriae; 10 – arcus palmaris profundus; 11 – a. radialis; 12 – a. recurrens radialis; 13 – a. collateralis radialis; 14 – a. profunda brachii; 15 – a. circumllexa humeri anterior; 16 – a. circumflexa humeri posterior

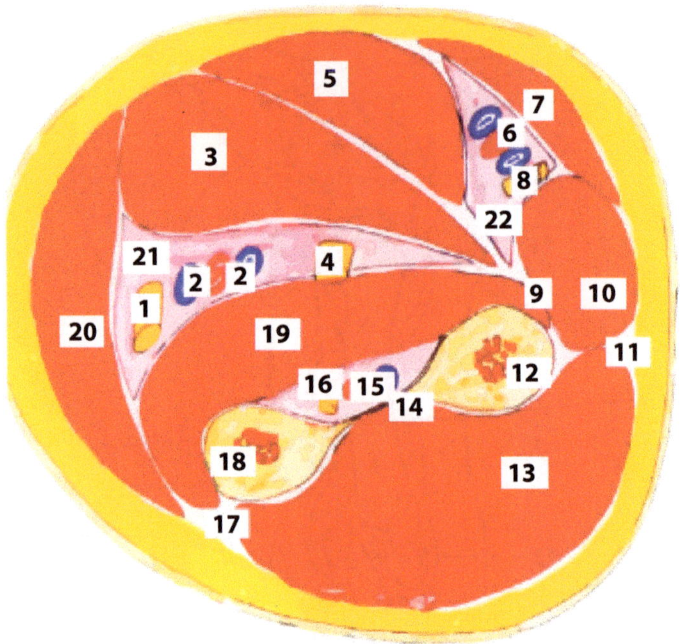

Figure 73 Crosscut of the left forearm in the middle third.
1 – n. ulnaris; 2 – a. et vv. ulnaris; 3 – m. flexor digitorum superficialis; 4 – n. medianus; 5 – m. flexor carpi radialis; 6 – a. et vv. radialis; 7 – m. brachioradialis; 8 – r. superficial n. radialis; 9 – septum intramusculare externum; 10 – extensor carpi radialis longus et brevis; 11 – septum intramusculare internum; 12 – radius; 13 – m. extensor digitorum; 14 – membrana interossea; 15 – vasa interosse anterior; 16 – n. interosseus anterior; 17 – septum intermusculare internum; 18 – ulna; 19 – m. flexor digitorum longus et m. flexor pollicis longus; 20 – m. flexor carpi ulnari

On the back side, two and a half fingers from the radius side are innervated by n. radialis, and two and a half fingers are innervated by n. ulnaris).(Figure 82–84 and 86)

The tendons of the fingers are immersed in a particular tendon sheath, secreting lubricant – synovial fluid. It is important to know that the tendon sheath of 1 and 5 fingers are longer. They penetrate on the forearm, where they merged with each other. The tendon sheath of 2, 3 and 4 fingers end in the palm of the hand. Therefore, inflammation of the tendon sheath 1 and 5 fingers is more dangerous because infection penetrates the forearm, where may occur deep abscess. Pus

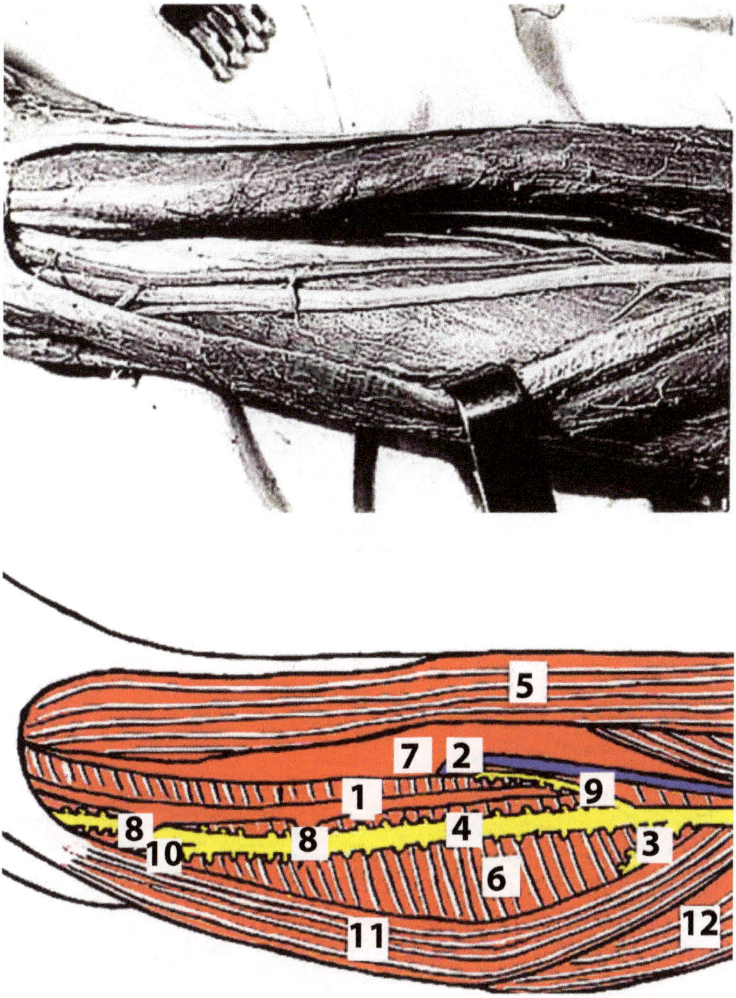

Figure 74 The front surface of the left forearm.
1 – a. ulnaris; 2 – v. ulnaris; 3 – r. muscularis n. ulnaris; 4 – n. ulnaris; 5 – m. flexor carpi ulnaris; 6 – m. flexor digitorum superficialis; 7 – m. flexor digitorum profundum; 8 – r. muscularis; 9 – r. communicans n. medialis; 10 – r. dorsalis n. ulnaris; 11 – r. palmaris n. ulnaris; 12 – m. palmaris longus; 13 – m. flexor carpi radialis; 14 – m. extensor carpi radialis brevis

in these cases is concentrated in a special space of Pirogov-Parona, located in the distal part of the forearm between the deep flexor of the fingers and the deeper lying square pronator. Incisions in these cases are made on the sides of the distal part of the forearm with the

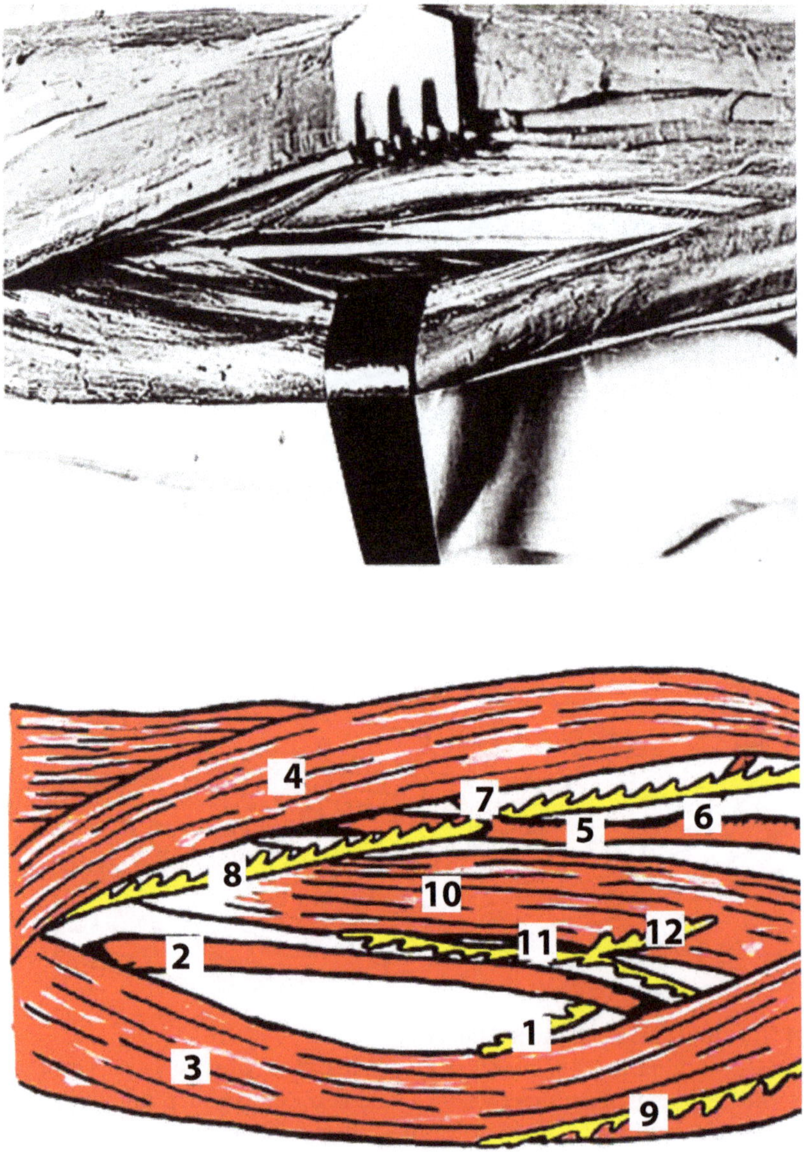

Figure 75 The front surface of the right forearm.
1 – r. superficialis; 2 – a. radialis; 3 – m. brachioradialis; 4 – m. flexor carpi radialis; 5 – a. ulnaris; 6 – a. recurrens ulnaris; 7 – r. muscularis a. ulnaris; 8 – n. ulnaris; 9 – n. musculocutaneus; 10 – m. flexor digitorum superficialis; 11 – n. medianus; 12 – r. anterior n. cutanei antebrachii medialis

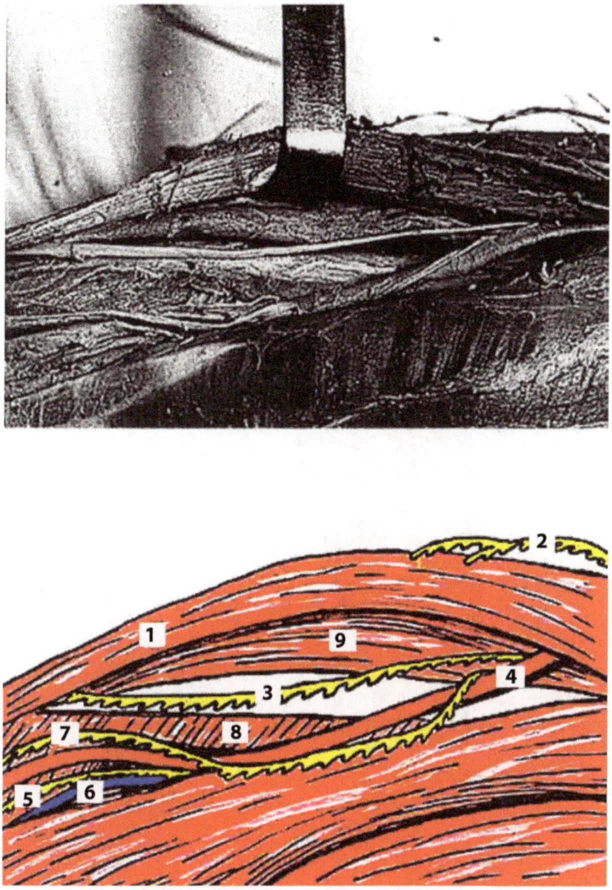

Figure 76 Topography of the upper third of the right forearm.
1 – m. brachioradialis; 2 – r. ulnaris n. cutanei antebrachii medialis; 3 – r. superficialis n. radialis; 4 – a. radialis; 5 – n. ulnaris; 6 – v. ulnaris; 7 – n. cutaneus antebrachii medialis; 8 – m. flexor digitorum superficialis; 9 – m. extensor carpi radialis brevis

institution hrough them through gauze wipes. As a result, there will be a good drainage of the entire purulent cavity.

Among surgeons in respect of purulent tendovaginitis is widely believed: "when tendovaginitis 1 and 5 fingers is responsible for the whole hand; when tendovaginitis 2, 3 and 4 fingers – only each individual finger".

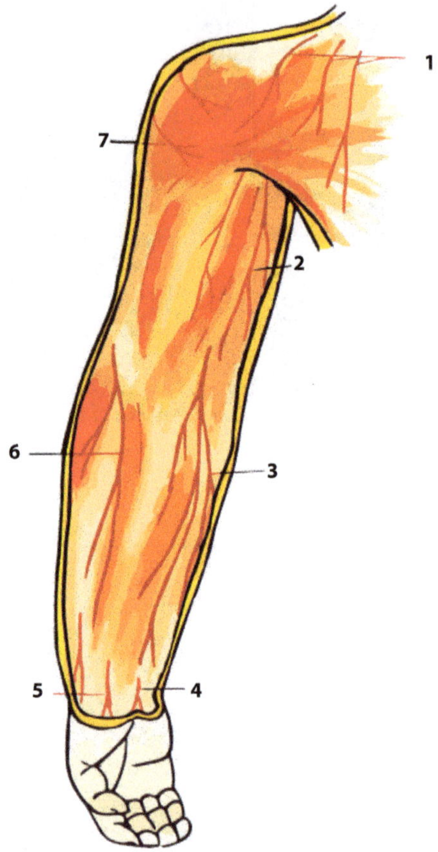

Figure 77 Subcutaneous nerves of the upper limb. Front view.
1 – nn. supraclaviculares media; 2 – n. cutaneus brachii medialis; 3 – n. cutaneus antebrahii medialis; 4 – r. palmaris n. ulnaris; 5 – r. palmaris n. mediani; 6 – n. cutaneus antebrahii lateralis; 7 – n. cutaneus brachii lateralis

Surgical Anatomy of Congenital Malformations of the Upper Limb

Syndactyly is a deformation, characterized by complete or incomplete connection of several fingers on the hand as a result of the delay of the normal process of splitting segments. There is a division of syndactyly into the forms: simple and complex, complete or incomplete. The simple form has four groups: skin, membranous, terminal and bone. The complex form occurs, when other malformations of the limbs are combined.

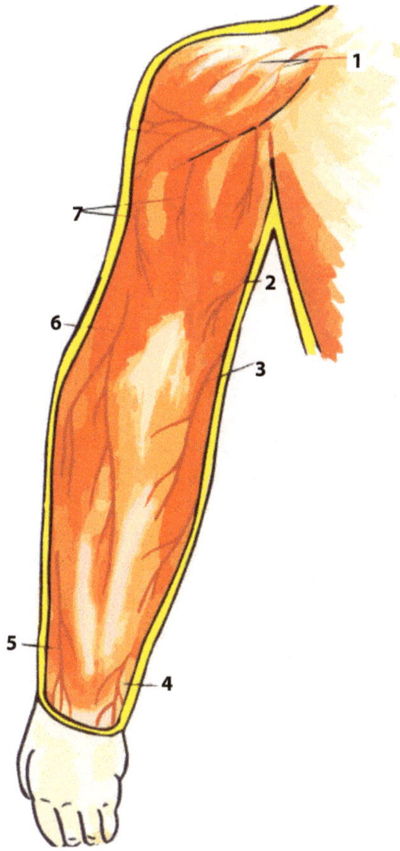

Figure 78 Subcutaneous nerves of the upper limb. Back view.
1 – nn. supraclaviculares media; 2 – n. cutaneus brachii medialis; 3 – n. cutaneus antebrachii medialis; 4 – r. palmaris n. ulnaris; 5 – r. palmaris n. median; 6 – n. cutaneus antebrachii lateralis; 7 – n. cutaneus brachii lateralis

Polydactyly is a deviation when a person has additional fingers, it can be combined with other developmental abnormalities. The additional fingers are more often located on the radial or ulnar edge of the hand, they can approach the structure of the normal fingers or represent rudimentary appendages.

Congenital clubhand – the persistent deviation of the hand in the direction of the missing or underdeveloped forearm bone – radial or elbow strabismus. It often combined with other abnormalities of development.

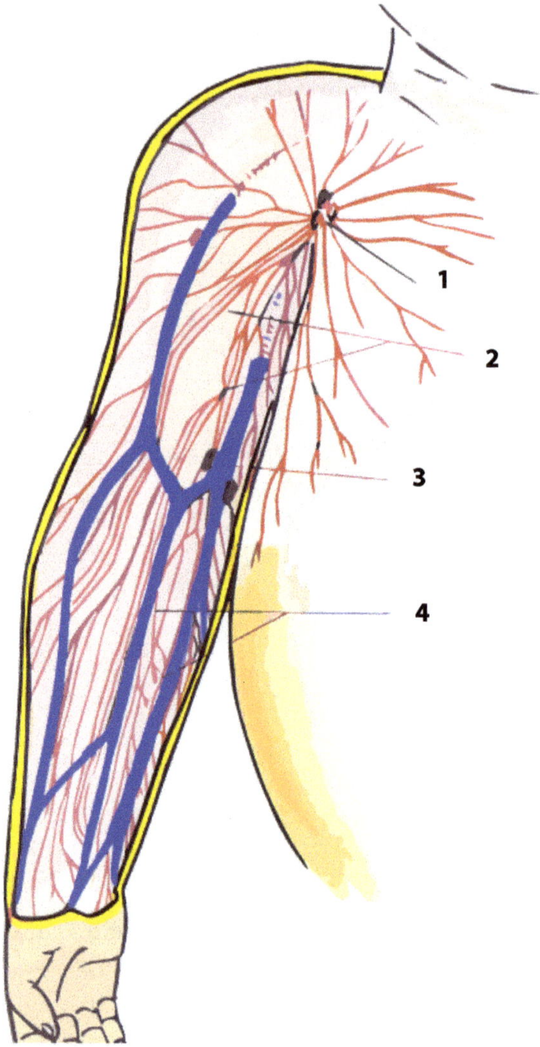

Figure 79 Subcutaneous lymphatic and venous vessels of the upper limb.
1 – nodi lymphatici axillares superficialis; 2 – vasa lymphatica brachii; 3 – nodi lymphalici cubitales; 4 – vasa lymphatica antebrahii

Congenital radiocubital synostosis – the deformation, manifested by the fixed position of the distal forearm in varying degrees of pronation.

Congenital high scapula (Sprengel's deformity) – a malformation of the shoulder girdle, characterized by changes in the symmetry of the shoulder blades, the angle of the shoulder blade, the degree of

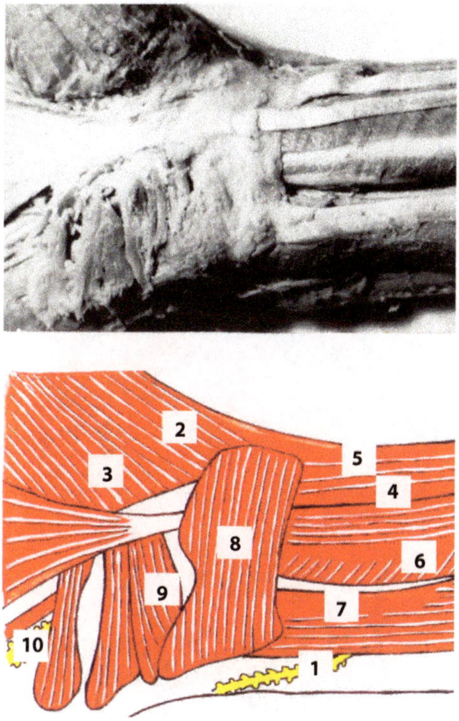

Figure 80 Palm surface of the wrist joint.
1 – r. dorsalis n. ulnaris; 2 – m. abductor policis brevis; 3 – m. flexor policis brevis; 4 – m. palmaris longus; 5 – m. flexor carpi radialis; 6 – m. flexor digitorum superficialis; 7 – m. flexor carpi ulnaris; 8 – m. abductor digiti minimi; 9 – m. flexor digili minimi brevis; 10 – rr. superficiales n. ulnaris

diversion of the shoulder, as well as the hypotrophy of the romobovid, trapezoid muscles and muscles of the shoulder girdle.

Winged scapula – deformation of the shoulder blade due to underdevelopment or lack of romobovid, trapezoidal and anterior serratus muscles (congenital), develops as a consequence of polio and paralysis of the above muscles.

Pathotopography of the Upper Limbs

Osteomyelitis is an inflammation of the bone marrow, which usually affects all elements of the bone (periosteum, spongy and compact

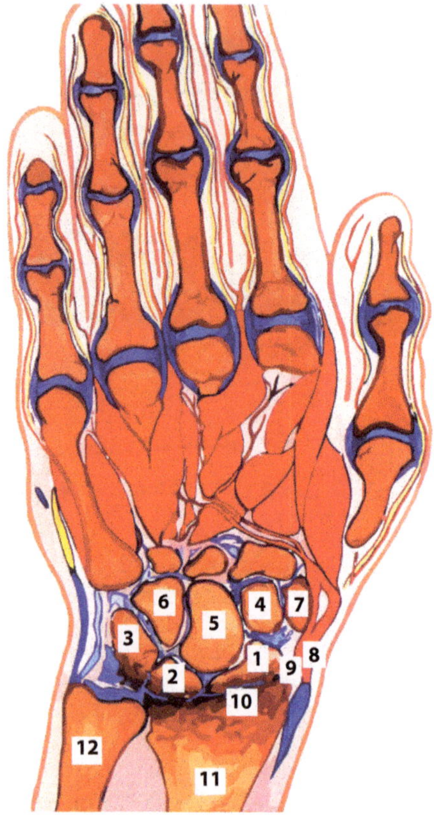

Figure 81 Frontal section of the hand.
1 – os scaphoideum; 2 – os lunatum; 3 – os triquetrum; 4 – os trapesoideum; 5 – os capitatum; 6 – os hamarum; 7 – os trapezium; 8 – a. ulnaris; 9 – anatomical snuffbox; 10 – rheumatic swelling of the radius; 11 – radius; 12 – ulna

substance). More often affects femur and humerus, bones of the lower leg, vertebrae, mandibular joints and upper jaw. Men suffer from osteomyelitis more often than women. Depending on the etiology of osteomyelitis, it is divided into nonspecific and specific (tubercular, syphilitic, brucellosis, etc.); post-traumatic, hematogenous, postoperative, contact. Sequestrants can form in chronic cases of osteomyelitis. A sequestrum is a bone site that freely floats in the cavity of the medullary canal, which has been torn off from a compact or spongy substance due to purulent fusion. It is one of the signs, in determining which it is possible to say with certainty that the patient has osteomyelitis. At

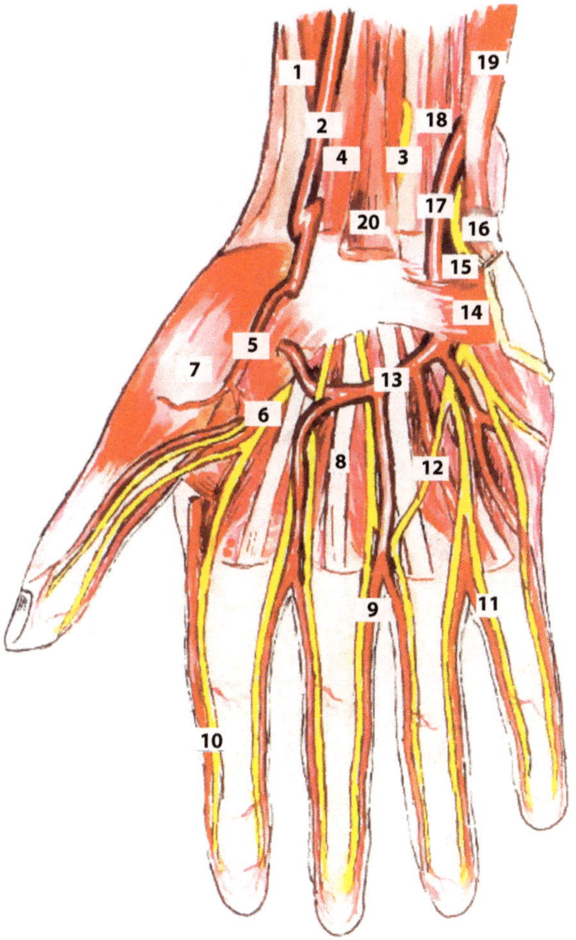

Figure 82 The topography of the surface palm formations.
1 – tendo brachioradialis; 2 – a. radialis; 3 – n. medianus; 4 – m. flexor carpi radialis; 5 – r. palmaris a. ulnaris superficialis; 6 – r. digitalis communis; 7 – m. extensor et adductor pollicis; 8 – tendo m. flexor pollicis superficialis; 9 – a. digitalis palmaris communis; 10 – aa. et nn. digitales palmares propriae; 11 - nn. digitales palmares communes; 12 – anastomosis between the branches of the median and ulnar nerve; 13 – arcus palmaris superficialis; 14 – m. palmaris brevis; 15 – r. a. ulnaris profunda; 16 – os pisiforme; 17 – a. ulnaris; 18 – m. flexor digitorum superficialis; 19 – m. flexor carpi ulnaris; 20 – m. palmaris longus

the formed fistulous course the sequestrum can be allocated from it together with pus. In children, resorption (resorption) of the formed sequester in the acute phase of the disease can occur. When passing

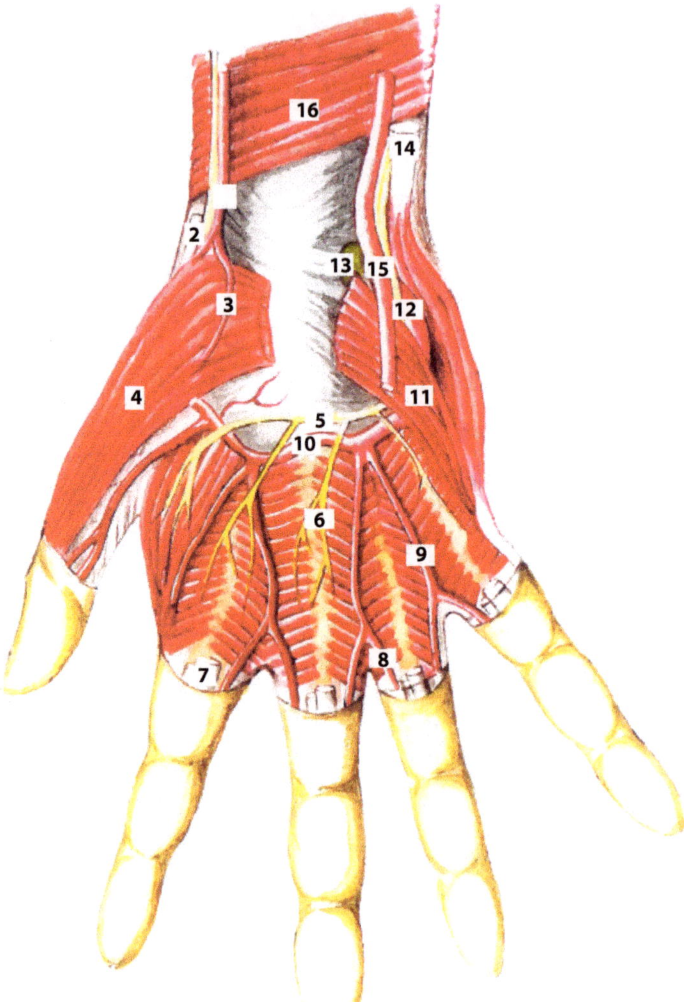

Figure 83 The deep palmar structures.
1 – a. radialis; 2 – tendo m. abductor pollicis longus; 3 – r. palmaris a. radialis superficialis; 4 – m. abductus pollicis longus; 5 – r. profundus n. ulnari; 6 – The branches of the ulnar nerve to the interosseous muscles; 7 – tendo in. flexor digitorum; 8 – a. digitorum propria; 9 – aa. metacarpis; 10 – r. palmaris profundus; 11 – m. eminentii digiti minimi; 12 – r. profundus a. et n. ulnares; 13 – os psiforme; 14 – tendo m. flexor carpi ulnaris; 15 – a. ulnaris; 16 – m. pronator quadratus

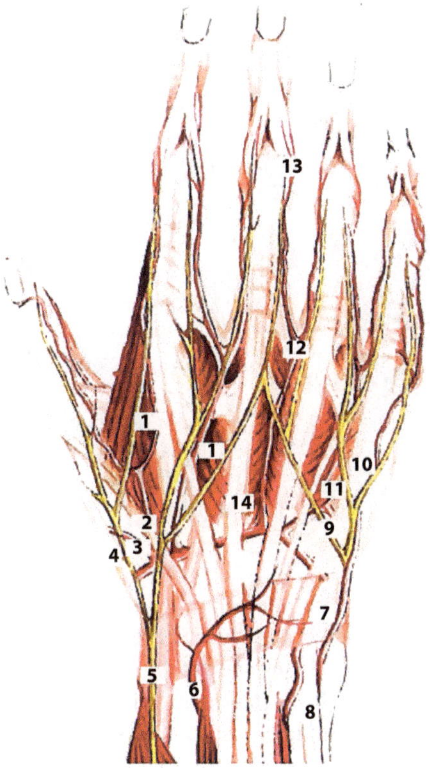

Figure 84 Posterior surface of the pulm.
1 – mm. interossi dorsales; 2 – tendo m. extensor pollicis longus; 3 – a. ulnaris «in the snuffbox»; 4 – tendo m. extensor pollicis brevis; 5 – r. superficialis n. ulnaris; 6 – a. palmaris interossea; 7 – ligament that holds the tendons of the extensor; 8 – m. extensor carpi; 9 – r. dorsalis n. ulnaris; 10 – tendo m. extensor digiti minimi; 11 – a. metacarpi dorsalis; 12 – aa. digitorum dorsales; 13 – rr. dorsales aa. digitales palmares propriae; 14 – tendo m. extensor digitorum

into the chronic course, a protective capsule forms around it, which prevents both resorption and attachment to a healthy bone. With age, the ability of sequestrants to self-dissolving decreases.

Pathotopography (Figures 90–92). Once the bone tissue is affected by microorganisms, leukocytes migrate to the source of infection, they secrete lytic enzymes that break down the bone.

In this figure, the process of sequestration is localized within the humerus. To the sequestration of the bone has led the spread of pus on the blood vessels, thus forming the basis for chronic infection.

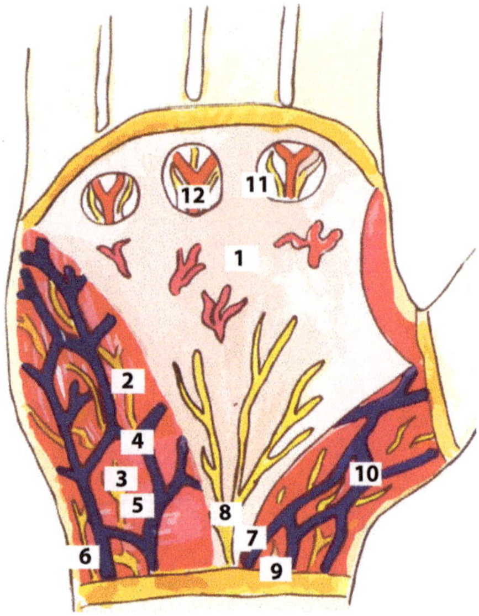

Figure 85 Subcutaneous structures of the palm.
1 – aponeurosis palmaris; 2 – branch of the ulnar nerve to the 5th finger; 3 – r. palmaris n. ulnaris superiicialis; 4 – vv. superficiales eminentis digiti minimi; 5 – m. palmaris brevis; 6 – r. n. cutaneus antebrachii medialis; 7 – tendo m. palmaris longus; 8 – r. palmaris n. mediani; 9 – r. n. cutaneus antebraehii lateralis; 10 – vv. superficiales eminentis pollici; 11 – foramen comissura; 12 – aa. et nn. digitorum

The upper border of the anterior and posterior regions of the shoulder is the conditional line connecting the lower extremities of the large thoracic and broadest muscles on the shoulder, the lower one is the line passing through 2 transverse fingers above the epicondyle of the humerus.

In relation to the osteomyelitis cavity, there are three possible variants of the location of the sequester: when it is entirely in the cavity, partially exits the cavity and completely outside the bone cavity in the soft tissues. When an extraosseous sequestrum is located within the surrounding soft tissues of the humerus, osteomyelitis can lead to the displacement of the shoulder muscles (coracoid-brachial, biceps arm and shoulder muscles, triceps brachium muscle and elbow muscle) in any direction. The size of the sequestration can be different depending on the depth of damage to the bone tissue. Large-sized sequesters can have a compressive effect on the contents of the shoulder cannula,

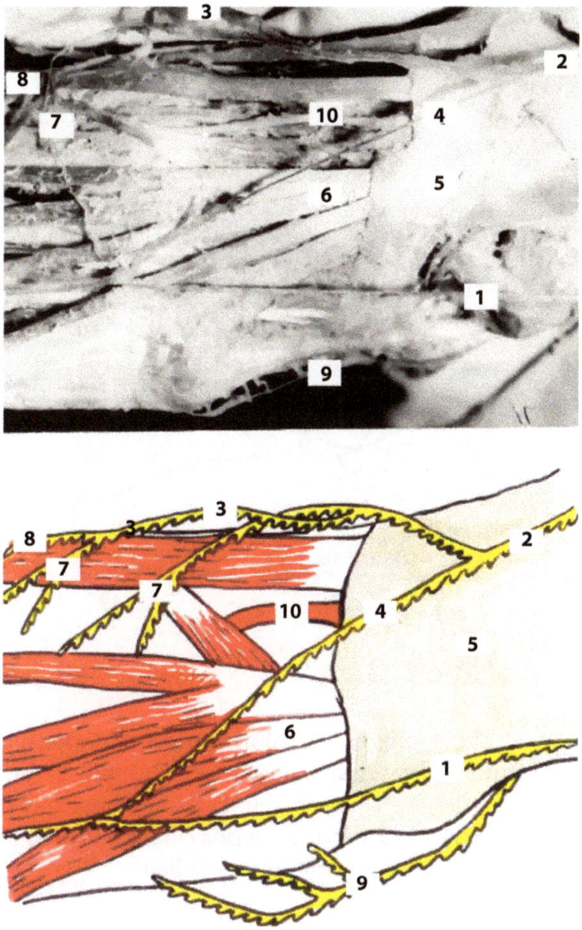

Figure 86 Topography of the palmar surface of the wrist joint.
1 – n. plantaris medialis; 2 – n. plantaris lateralis; 3 – r. superficialis; 4 – n. digitalis palmaris communicans; 5 – aponeurosis plantaris; 6 – m. flexor digitorum brevis; 7 – r. cutanei; 8 – m. abductor digiti minimi; 9 – rr. cutanei (n. plantaris medialis); 10 – a. plantaris lateralis

the space between the humerus and the heads of the muscle triceps brachial; the location of the radial nerve, deep arteries and veins of the shoulder and the contents of the medial sulcus of the biceps muscle (a longitudinal depression on the inner surface of the shoulder between the medial head of the triceps muscle behind, the brachial and biceps muscles in the front, the neurovascular bundle passes in the furrow),

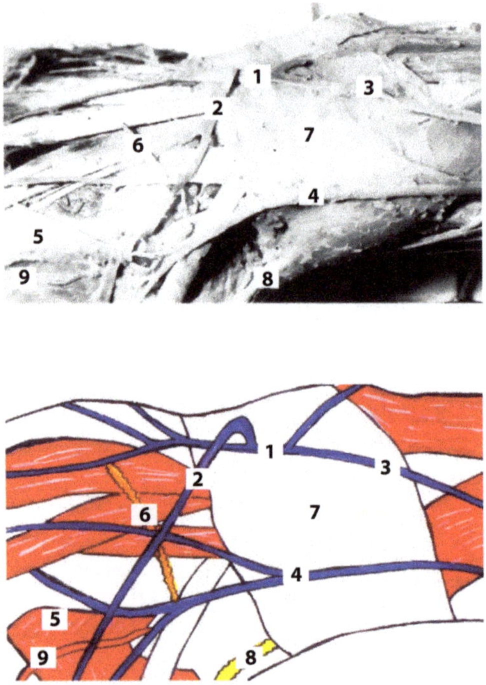

Figure 87 The topography of the veins on the palmar surface of the wrist joint. 1 – vv. carpaliae posteriores; 2 – rete venosum dorsale; 3 – v. capitis; 4 – m. abductor indicis; 6 mm. interossei dorsales; 7 – Link between the surface branch of the radial nerve at the back of the branches of the ulnar nerve; 8 – retina culum flexorum manus; 9 – r. superticialis n. radialis; 10 – rr. carpaliae posteriores a. radialis

resulting in impaired blood supply and innervation muscles. The subcutaneous fatty tissue is deformed and has different thicknesses in different areas.

Traumatic edema of the radial bone is an excessive accumulation of fluid in the organs, extracellular tissue spaces of the body. Artificial edema and swelling, located mainly on the limbs, can be obtained in two ways:

1. The limb is pulled by a bandage, lace, towel, elastic band, etc.;
2. Weak piercing was performed systemically for a long time on a certain part of the body.

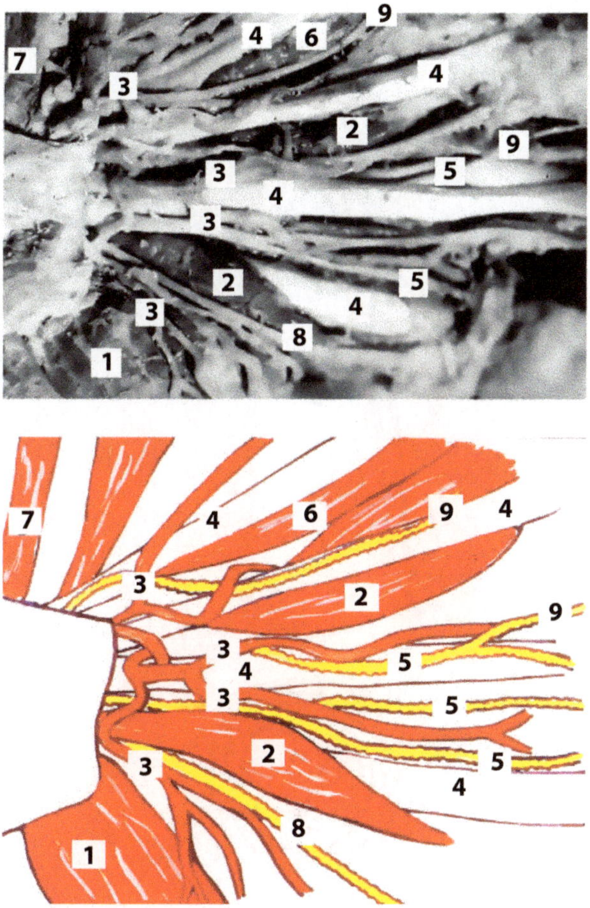

Figure 88 Topography of the palm.
1 – m. flexor policis brevis; 2 – mm. lumbricales; 3 – a. digitalis palmaris communis; 4 – tendines m. flexoris digitorum superficialis; 5 – n. digitalis palmaris profundus; 6 – m. flexor digiti minimi brevis; 7 – m. palmaris brevis; 8 – r. superficialis n. ulnaris; 9 – r. nervalis m. lumbricali

Pathotopography (Figure 93). The preferred localization of traumatic edema is the dorsal surface of the hands, more often the left, the posterior surface of the feet and the anterior surface of the shins. Edema, as a rule, does not extend to the thumbs of the feet and hands due to the considerable soreness of their effleurage and the lack of a layer of soft tissue.

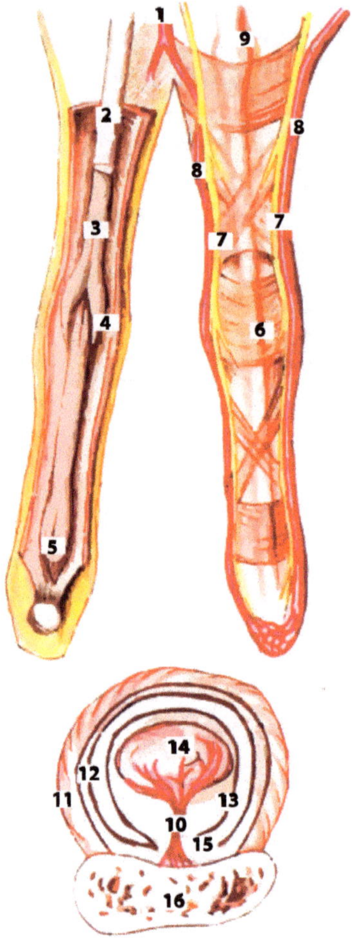

Figure 89 Fingers. l – aa. digitales palmares communes; 2 – tendo m. flexor digitorum profundus; 3 – tendo m. flexor digitorum superficialis; 4 – place of attachment of m. flexor digitorum superficialis; 5 – place of attachment of m. flexor digitorum profundus; 6 – lig. collaterale; 7 – n. digitalis palmaris proprium; 8 – a. digitalis palmaris proprium; 9 – the proximal end of the synovial vagina of finger tendons; 10 – vessels supplying the tendons; 11 – tunica fibrosa tendines; 12 – the parietal leaf of the synovial vagina of tendon; 13 – the visceral leaf of the synovial vagina of tendon; 14 – tendes mm. flexoribus (in cross section) 15 – the mesentery tendon; 16 – phalanx (cut)

The consequence is a gradual development of persistent edema, in this case, traumatic edema of the radius. The brush swells, loses its contours, the skin is infiltrated, reddish-cyanotic color. Movement

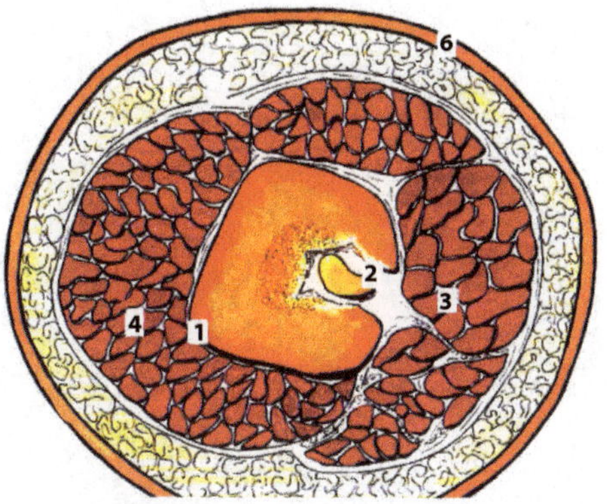

Figure 90 Cross section of the shoulder. Osteomyelitis of humerus.
The osteomyelitis of the humerus (1) without sequestration in the surrounding soft tissues, the pathological process is localized within the humerus (2). The contents of the cannula (3) and the contents of the medial sulcus of the biceps (4) are not visualized. 5 – subcutaneous fatty tissue; 6 – skin

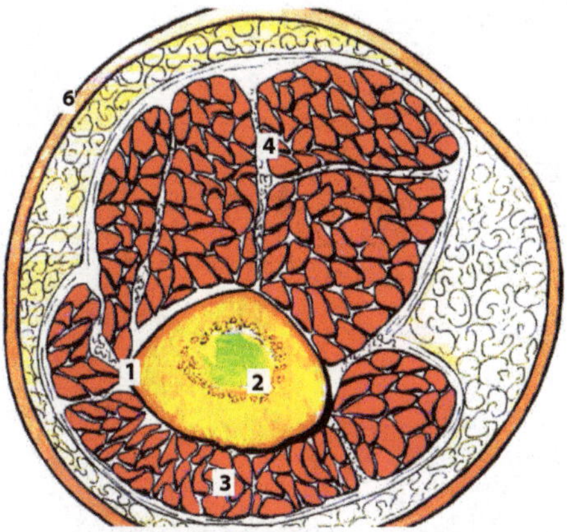

Figure 91 Cross section of the shoulder. Osteomyelitis of the humerus.
Osteomyelitis of the humerus (1) with an extra-osseous sequestrum within the surrounding soft tissues (2). The contents of the cannula (3) and the contents of the medial sulcus of the biceps (4) are not visualized. 5 – subcutaneous fat; 6 – skin

in the joint can be limited without particular soreness. Changes in bone tissue are not detected. By skin inspection through the magnifying glass can sometimes be detected in various places, mainly behind the ankles, retraction of punctate scars, which are more noticeable in

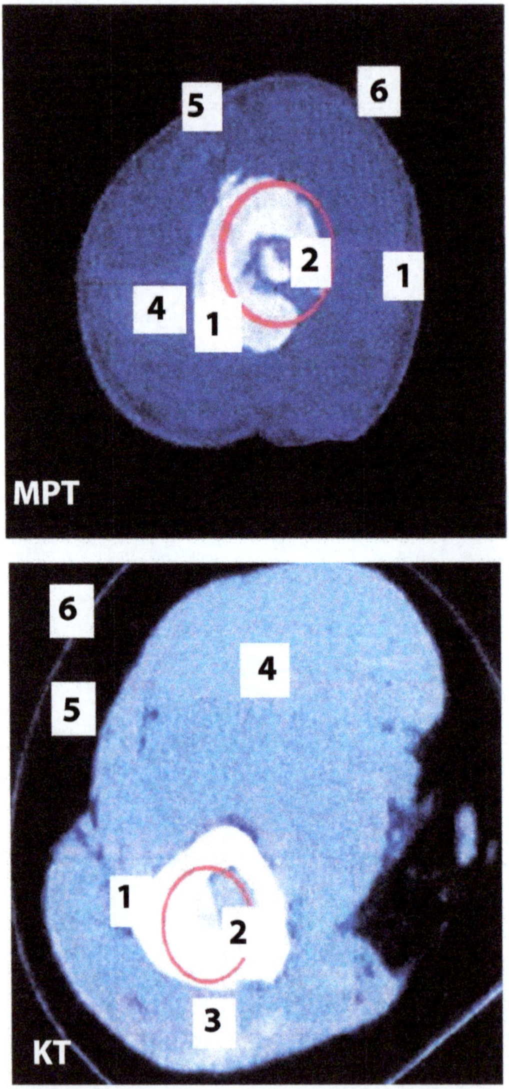

Figure 92 Pathotopographic anatomy of the transverse section of the shoulder. 1 – osteomyelitis of the humerus; 2 – soft tissues; 3 – contents of the cannula; 4 – the content of the medial sulcus of the biceps muscle; 5 – subcutaneous fatty tissue; 6 – skin

lateral illumination and pressure by the finger on the skin next to this, in order to displace it and detect retraction.

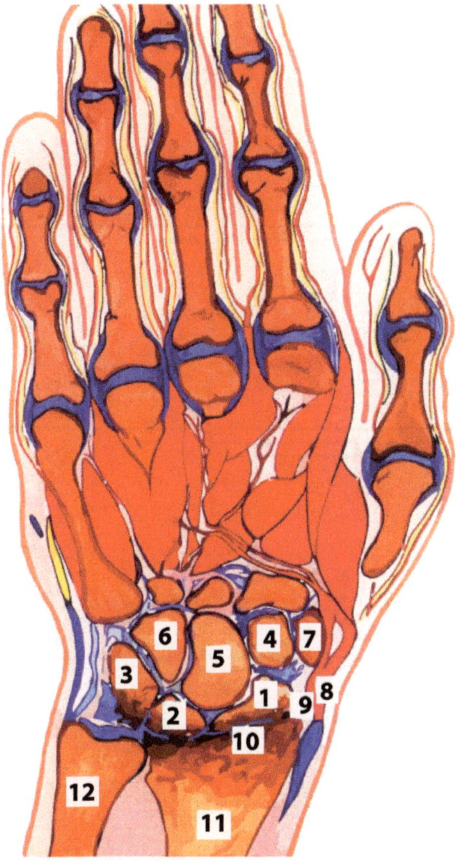

Figure 93 Frontal section of the hand.
1 – os scaphoideum; 2 – os lunatum; 3 – os triquetrum; 4 – os trapesoideum; 5 – os capitatum; 6 – os hamarum; 7 – os trapezium; 8 – a. ulnaris; 9 – anatomical snuffbox; 10 – traumatic edema of the radius; 11 – radius; 12 – ulna

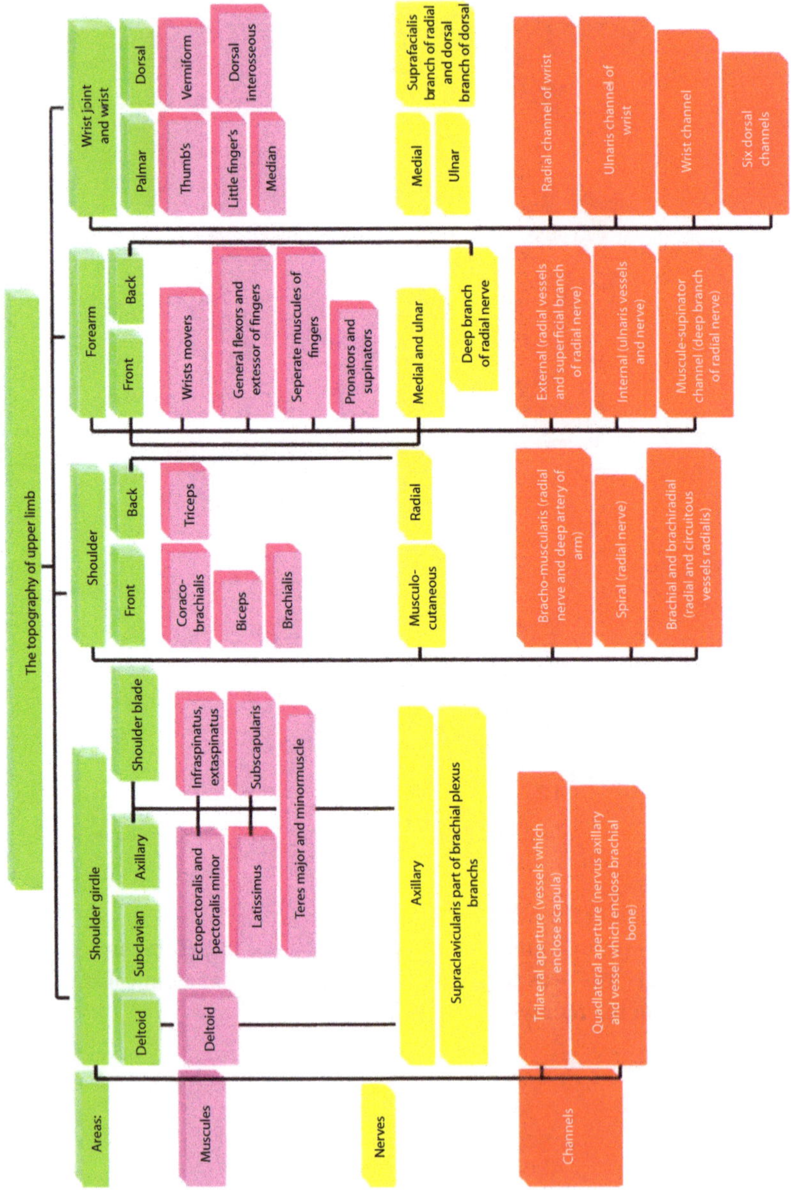

Topographic Anatomy of Lower Limbs

The lower limb is subdivided into the gluteal region, thigh, calf, foot.

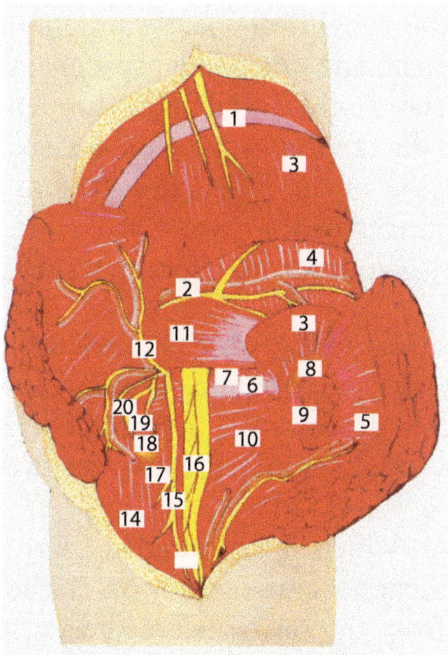

Figure 94 Gluteal region.
1 – crista ossis ilii; 2 – a. et n. gluteus superior; 3 – m. gluteus medius; 4 – m. gluteus minimus; 5 – m. gluteus maximus; 6 – tendo m. obturatorius internus; 7 – mm. gemellus; 8 – trochanter major; 9 – bursa trochanterica; 10 – m. quadratus femoris; 11 – m. piriformis; 12 – a. et n. gluteus inferior; 13 – m. biceps femoris; 14 – m. semitendinosus; 15 – n. eutaneus femoris posterior; 16 – n. ischiadicus; 17 – tendo m. semimembranosus; 18 – tuber ischiadicus; 19 – lig. sacrotuberale; 20 – a. et n. pudendus internus

Gluteal Region

Layers

- the superficial layer of these muscles: m. gluteus maximus, lamina profunda fasciae gluteal separates m. gluteus maximus from deeper fat, stratum adiposum;
- the middle layer these muscles: m. gluteus medius; m. piriformis; m. gemellus superior et inferior; there are tendo m. obturatorii interni; m. quadratus femoris between them;
- deep layer of muscle: m. gluteus minimus; m. obturatorius externus (Figure 94).

Cellular spaces: the subgluteal space communicates through the proper hole and subpiriforme side cellular spaces of the pelvis; through a small hole with a sciatic ischiorectal fossa; through the lacuna under the proximal part of the tendon of the gluteus maximus with the lateral and anterior region of the thigh; along the sciatic nerve – with fiber posterior thigh fascial bed.

The next structures go out through foramen subpiriforme:

- n. ischiadicus (Figure 95), which is medially to n. cutaneus femoris posterior, a. comitans n. ischiadici. The sciatic nerve is directed downward – in front of his upper gemellus tendon internal sphincter, the lower twin and square muscles.
- glutea inferior, which is surrounded by the same name veins and branches n. gluteus inferior. The bundle is opening inwards from the sciatic nerve and posterior femoral cutaneous nerve. Coming out of the hole, arteries and nerves break down the branches penetrating into the interior of the gluteus maximus and piriformis.
- pudendal neurovascular bundle (a. et v. pudendae internae and n. pudendus) is located in the most medial part of foramen subpiriforme.

Suprapiriformis hole (foramen suprapiriformis). There is a., v. et n. gluteus superior passing through the hole.

Subpiriforme hole (foramen infrapiriforme). There are a., v. et n. gluteus inferior (Figures 96–98), n. ischiadicus, a. pudenda interna et v. pudenda interna, n. pudendus and n. cutaneus femoris posterior passing through this hole.

Femur

The hip includes three groups of muscles: flexors, extensors and adductors. Each group is contained in a special fascial case. Topographically they distinguish front, back and medial muscle groups.

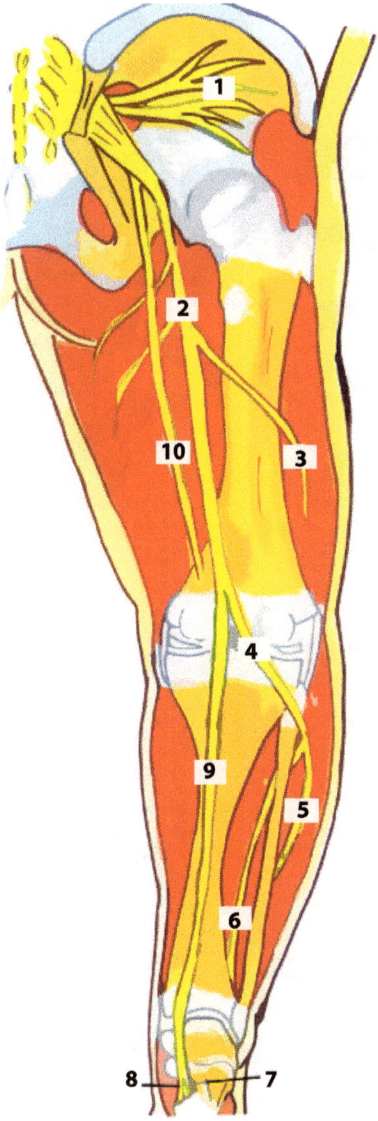

Figure 95 The sciatic nerve and its branches. Back view.
1 – n. gluteus superior; 2 – n. schiadicus; 3 – r. muscularis n. ischiadici; 4 – n. peroneus communis; 5 – n. peroneus superficialis; 6 – n. peroneus profundus; 7 – n. plantaris lateralis; 8 – n. plantaris medialis; 9 – n. tibialis; 10 – n. cutaneus femoris posterior

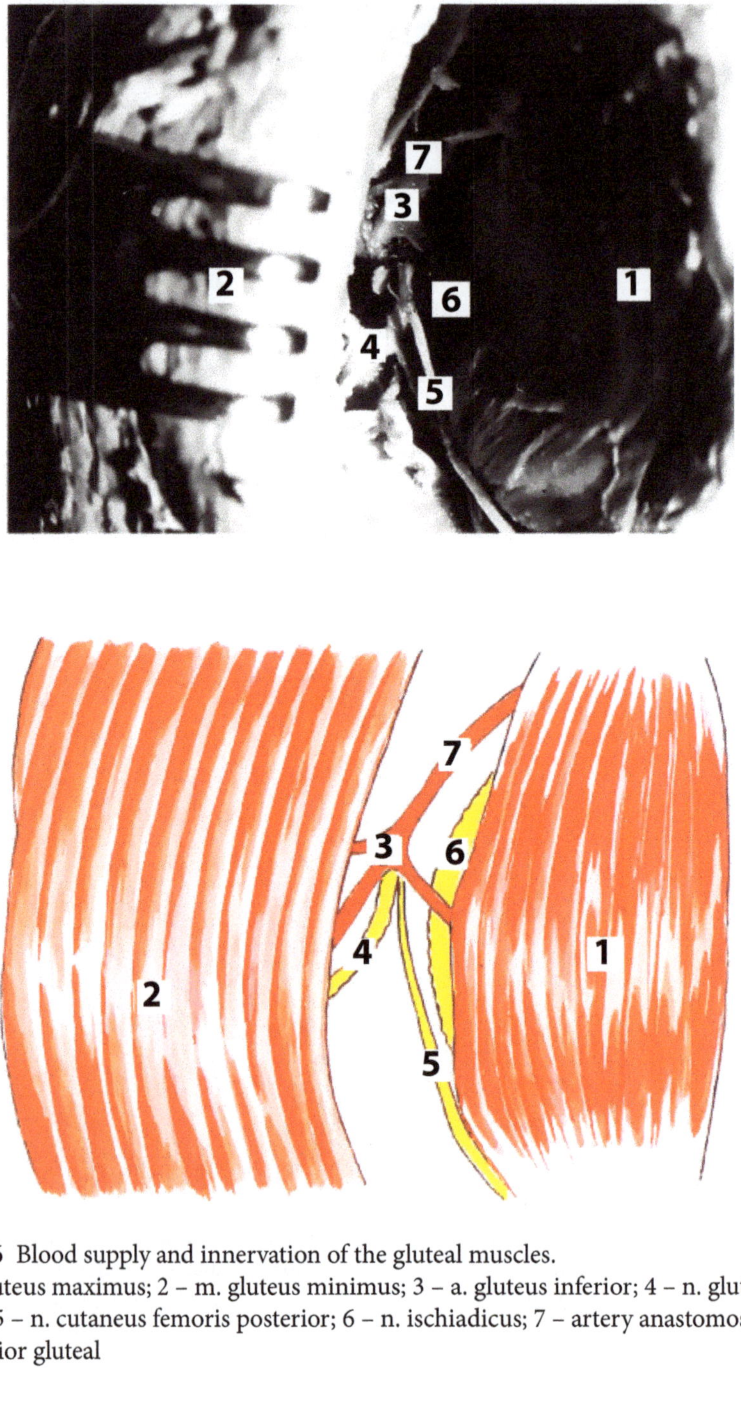

Figure 96 Blood supply and innervation of the gluteal muscles.
1 – m. gluteus maximus; 2 – m. gluteus minimus; 3 – a. gluteus inferior; 4 – n. gluteus inferior; 5 – n. cutaneus femoris posterior; 6 – n. ischiadicus; 7 – artery anastomosis with the superior gluteal

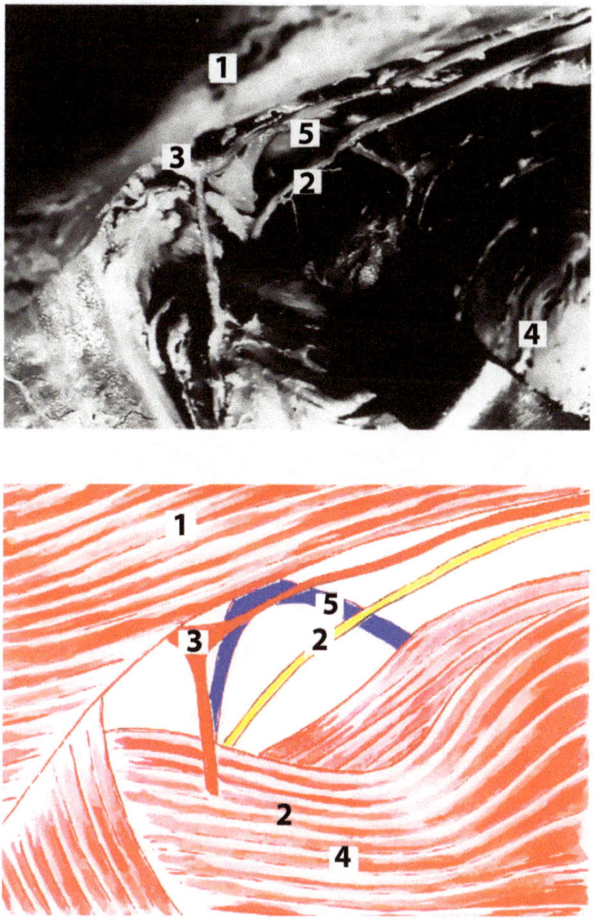

Figure 97 Blood supply and innervation of the gluteal muscles.
1 – m. gluteus maximus; 2 – n. gluteus superior; 3 – a. gluteus superior; 4 m. gluteus medius; 5 – v. gluteus superior

The front group includes the quadriceps femoris, m. quadriceps femoris (Figure 99). Its heads: rectus femoris, m. rectus femoris, broad internal muscles (m. vastus medialis); wide external muscle (m. vastus lateralis) and vastus intermediate (m. vastus intermedius). Intermediate and rectus muscles lie in the middle; the other two heads are on each side. Rectus covers intermediate front. M. quadriceps femoris extends the shin, and it contributes to the rectus and hip flexion (Figures 100 and 101).

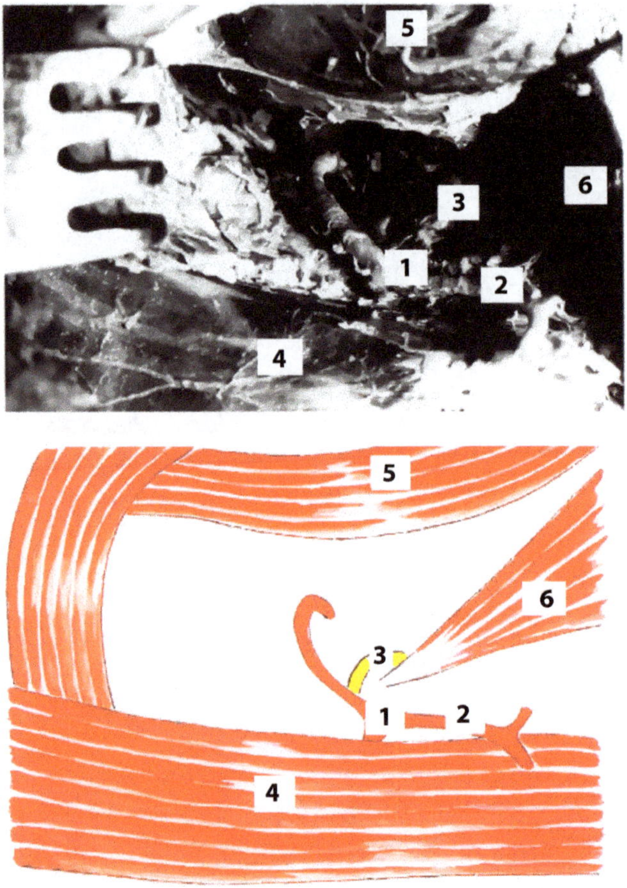

Figure 98 The topography of the gluteal region.
1 – truncus posterior a. iliacae internae; 2 – a. gluteus superior; 3 – n. gluteus superior; 4 – m. gluteus maximus; 5 – m. gluteus minimus; 6 – m. piriformis

The posterior group includes three flexors: two-headed thigh muscle, m. biceps femoris; semitendinosus (m. semitendinosus) and semimembranous muscle (m. semimembranosus). M. biceps femoris is located outside, medially – the other two muscles. All three muscle flex shin (Figures 102 and 103).

The medial group includes five adductors of thigh (Figures 100–102) – in the MR and CT scans you can see the contents of the leading canals (19) bounded by the front – internal vastus (16), posterior

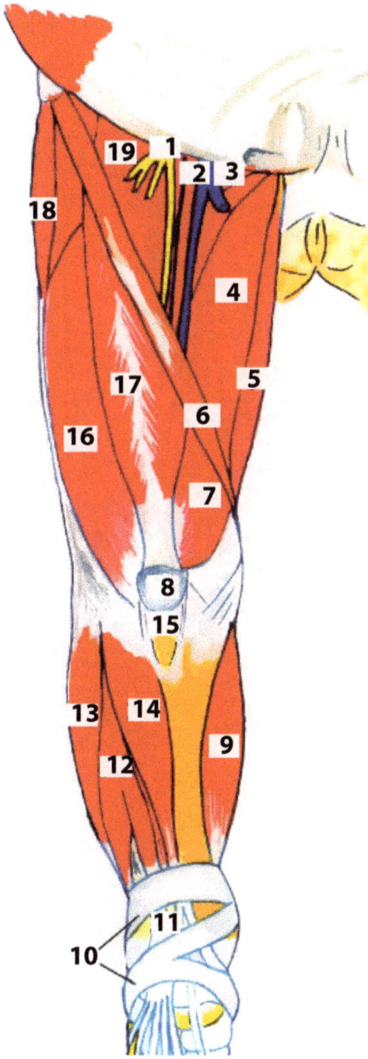

Figure 99 Muscles of the lower limb. Front view.
1 – n. femoralis; 2 – a. et v. femoralis; 3 – m. pectineum; 4 – m. adductor longus; 5 – m. gracilis; 6 – m. sartorius; 7 – m.vastus medialis; 8 – patella; 9 – m. triceps surae; 10 – retinaculum mm. extensorum superius et inferius; 11 – m. extensor digitorum longum. 12 – m. peroneus brevis; 13 – m. peroneus longus; 14 – m. tibialis anterior; 15 – lig. patellae; 16 – m. vastus lateralis; 17 – m. rectus femoris; 18 – m. tensor fasciae latae; 19 – m. iliopsoas

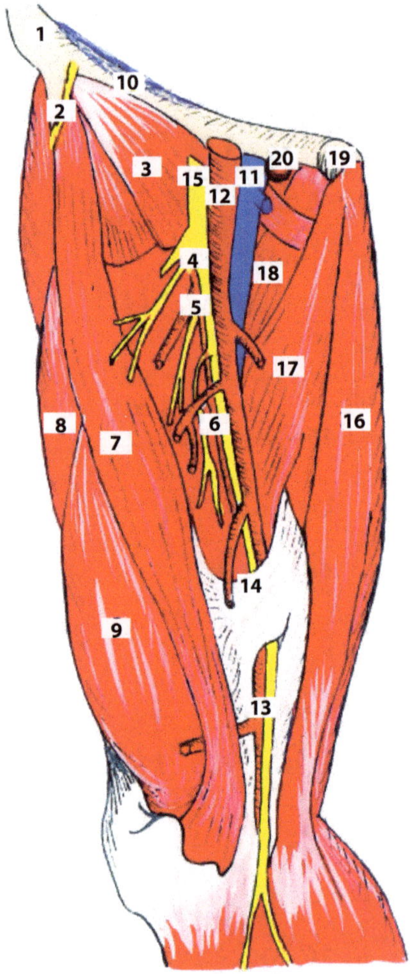

Figure 100 The femoral triangle.
1 – spina iliaca superior anterior; 2 – n. cutaneus femoris externus; 3 – m. iliopsoas; 4 – rr. culaneus et musculares n. femoralis; 5 – a. circullexa femoris externa; 6 – a. profunda femoris; 7 – m. sariorius; 8 – m. rectus femoris; 9 – m.vastus femoris iniernus; 10 – lig. inguinalis; 11 – v. femoralis; 12 – a. femoralis; 13 – n. saphenus et a. descedens; 14 – the lamina between the inner broad and adductor magnus muscle; 15 – n. femoralis; 16 – m. gracilis; 17 – m. adductor magnus; 18 – m.pectineus; 19 – tuberculum pubicum; 20 – Piragov's lymphatic node

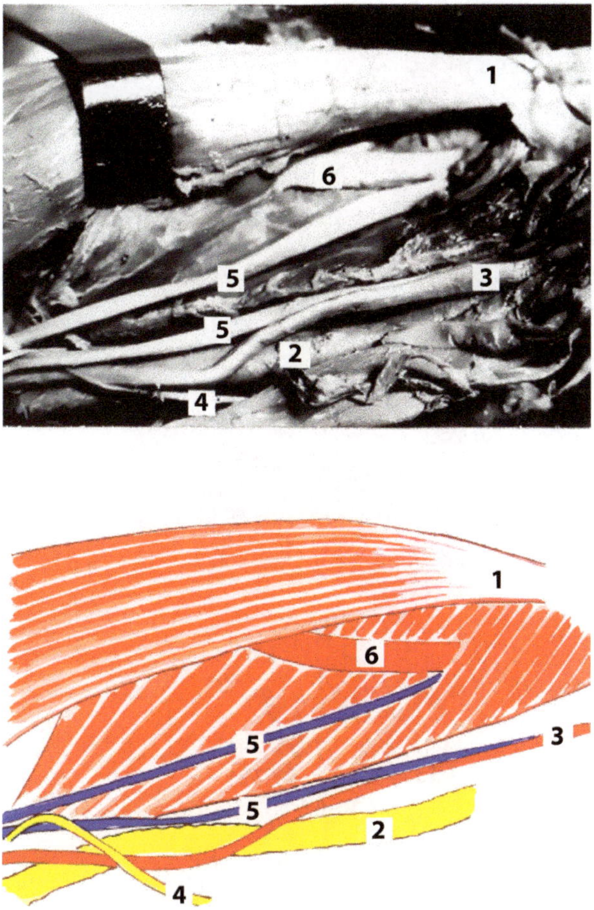

Figure 101 Blood supply and innervation of the back of the thigh.
1 – m. biceps femoris (caput longum); 2 – n. ischiadicus; 3 – a. profunda femoris; 4 – r. posterior n. obturatorii; 5 – v. profunda femoris; 6 – a. perforans

– adductor longus muscle (3) medially – intermuscular inner septum (2)): femoral vessels (2) and the saphenous nerve (20)):

1. m. adductor longus – adductor longus muscle;
2. m. adductor brevis – adductor brevis muscle lies several deeper and higher;
3. m. adductor magnus – adductor magnus muscle;
4. m. pectineus – a comb muscle;
5. m. gracilis – gentle muscle.

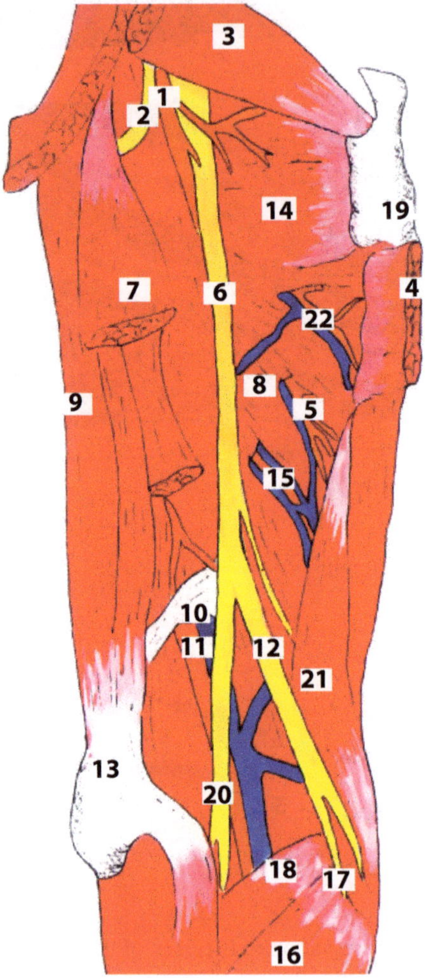

Figure 102 The posterior thigh area.
1 – a. glutea inferior; 2 – a. and n. pudendus; 3 – m. piriformis; 4 – m. gluteus maximus; 5 – a. perforanta (second); 6 – n. ischiadicus; 7 – envelopes shin muscles (the beginning); 8 – m. adduclor magnus; 9 – gentle muscle; 10 – the lower opening of the channel adductors; 11 – popliteal vessels; 12 – n. tibialis circumflexus; 13 – condylus femoris internus; 14 – m. quadrutus femoris; 15 – a. perforanta (third); 16 – m. gastrocnemius; 17 – n.cutaneus surae superficialis; 18 – m. plantaris; 19 – trochanter major; 20 – n. tibialis; 21 – m. biceps femoris; 22 – a. perforanta (first)

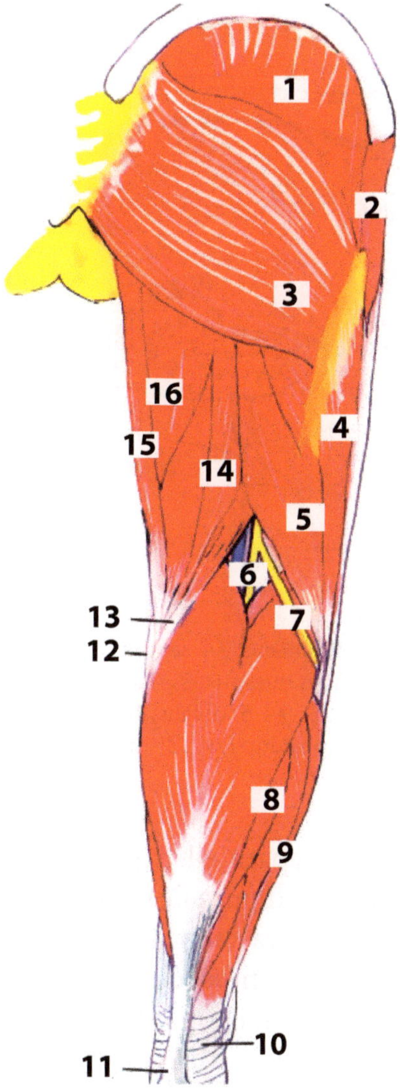

Figure 103 Muscles of the lower limb. Back view.
1 – m. gluteus medius; 2 – m. tensor fasciae latae; 3 – m. gluteus maximus; 4 – m. vastus lateralis; 5 – m. biceps femoris; 6 – v. poplitea et n. tibialis; 7 – n. peroneus communis; 8 – m. triceps surae; 9 – m. peroneus longus et brevis; 10 – retinaculum; 11 – tendo calcaneus; 12 – m. sartorius; 13 – m. semimembranosus; 14 – m. semitendinosus; 15 .m. gracilis; 16 – m. adductor magnus

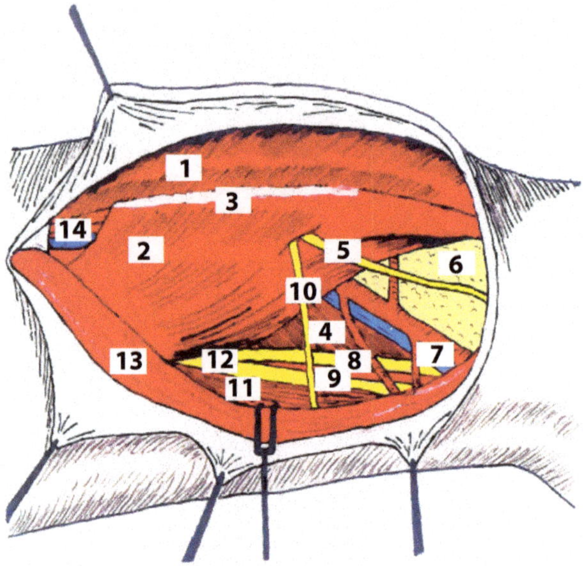

Figure 104 Channel adductors.
1 - m. sartorius; 2 – m. semimembranosus; 3 – septum intramusculare internum; 4 - m. vastus medialis; 5. n. cutaneus internus; 6 – fatty tissue of popliteal fossa; 7 – a. et v. popliteal; 8 – n. peroneus communis; 9 – n. tibialis; 10 – r. accessorius n. femoralis; 11 - lamina of wide muscle and hip adductors; 12 – n. ischiadicus; 13 – m. adductor magnus; 14 – a. et v. femoralis

At the top the hip is is separated from the abdomen by a special strong inguinal ligament (lig. inguinale). Under this ligament there are two lacunas; the medial side vascular lacuna (lacuna vasorum), with lateral – muscle lacuna (lacuna musculorum). A. and v. femoralis pass through the vascular lacuna, m. ileopsoas, the main hip flexor and the femoral nerve (n. femoralis) pass through the muscle.

Vascular lacuna (lacuna vasorum) is limited to the front by lig. inguinale, posterior – lig.pectineale, medially – lig. lacunare, laterally – arcus iliopectineus. Through the vascular lacuna may comes femoral hernia, forming a slit-shaped move – the femoral canal (canalis femoralis). Under normal conditions, it does not exist. The interior of its opening is called a femoral ring (anulus femoralis), referred to as the external subcutaneous course (hiatus saphenous). Lacunas hip iliac comb separates the arc (Figure 107).

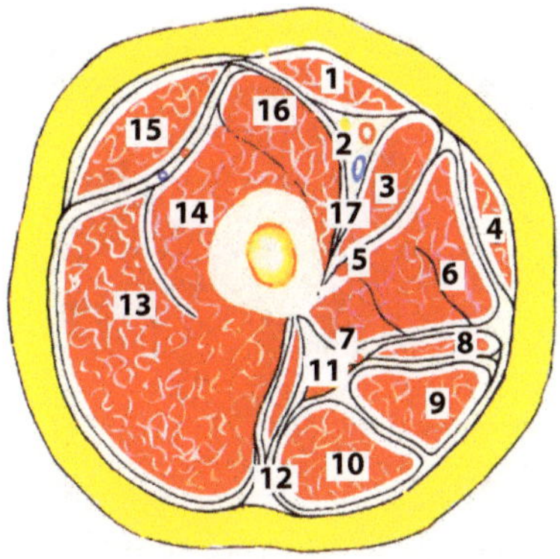

Figure 105 The cross-cut of the thigh.
1 – m. sartorius; 2 – internal intermuscular septum and femoral vessels; 3 – m. adductor longus; 4 – tender muscle; 5 – a. perforanta; 6 – m. adductor major; 7 – posterior intermuscular septum; 8 – m. semimembranosus; 9 – m. semitendinosus; 10 – m. biceps; ll – n. ischiadicus; 12 – external intermuscular septum; 13 – m. vastus externus; 14 – m. vastus medius; 15 – m. rectus femoris; 16 – r. vastus internus

Infringement of intestinal loop in the formation of a hernia usually occurs in the inner hole – femoral ring. Its boundaries are: the front lig. inguinale; posterior – lig. iliopubicum; medially – lig. lacunare and lateral – the femoral vein. This septum is the weakest: exiting hernia pushes vein outwards and it appears on the thigh below the inguinal ligament. Opening is closed leaf of the peritoneum, the transverse fascia, and most importantly a strong fibrous femoral septum (septum femorale), which prevents the development of a hernia. The hole is a lymph node of Pirogov-Rosenmüller.

Outside the opening of the femoral canal (Figure 108) – "subcutaneous course", hiatus saphenus, limited: top – the upper horn, cornu superius, below – the lower horn, cornu inferius, and outside – crescent edge, margo falciformis.

Femoral hernias are more common in women. This is due to the larger size of the femoral ring (average – 1.8 cm for men it is only 1.2 cm). The canalis has a triangular-slit-like shape, and is limited to the front by broad aponeurosis of thigh, aponeurosis latus, back – pectinate fascia, fascia pectinea, and outside – of the femoral vein. Hernial sac passes through the femoral ring and deflects femoral vein outwards, descends and appears on his hip under the skin through a

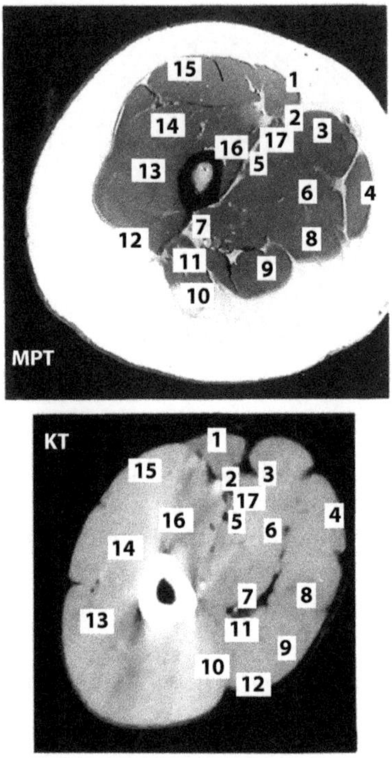

Figure 106 Cross-cut thigh (MRI and CT).

specified course – hiatus saphenus. This hole in a broad aponeurosis thigh is used to pass the great saphenous vein (v. saphena magna), which is infused into the femoral vein. It lies on the medial side of the thigh.

The femoral vein is located on the projection line of Ken, going from the middle of the inguinal ligament. During an artery ligation in the middle and lower third of the thigh roundabout circulation – at the expense of the descending branches of the medial and lateral arteries, envelopes femur.

Canals of Thigh

Adductor canal (canalis adductorius).

Septums: anterior-lateral – m. vastus medialis, posterior-medial – m. adductor magnus, front – lamina vastoadductoria, slung between the muscles and veiled m. sartorius.

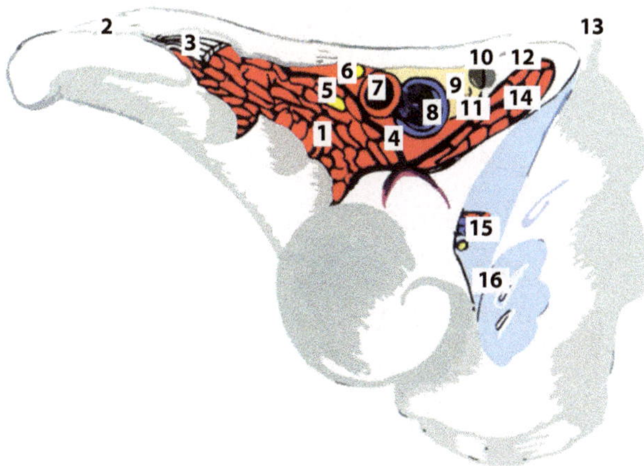

Figure 107 Vascular and neuromuscular lacunas.
1 – m. iliopsoas; 2 – spina iliaca superior anterior; 3 – lig. inguinalis; 4 – lig. iliopectineus; 5 – n. femoralis; 6 – the genital branch of the femoral nerve and hips; 7 – a. femoralis; 8 – v. femoralis; 9 – fat tissue of the inner ring of the femoral canal; 10 – nodus lymphaticus inguinalis profundus; 11 – fascia cristalis; 12 – lig. lacunara; 13 – tuberculum; 14 – m. pectineus; 15 – a. v. n. obturatorius; 16 – membrana obturatoria

Through upper foramen a. et v. femoralis, n. saphenus come in to canal; through the front exit a. genus descendens, n. saphenus; through the lower, located between the tendon m. adductor magnus and the femur, a. et v. femorales out in the popliteal fossa (Figure 109).

Adductor canal connects with popliteal fossa anterior thigh fascial bed.

The obturator canal (canalis obturatorius) is located in the obturator foramen. There are a., v. et n. obturatoriae in it. It is bounded above with sulcus obturatorius ossis pubis, below with membrana obturatoria, m. obturatorius int. et m. obturatorius ext.

Neoplastic Diseases: cartilaginous tumor (Figures 110 and 111) – on the CT scan it is visible cartilaginous tumor, cancer «in situ» (18), resulting in canal (19) is rendered, because it is not compressed limited pathological process, mesenchymal fibrosarcoma of thigh (Figures 112 and 113) – in the MR tomogram it is seen mesenchymal fibrosarcoma femur (17), compressing the outer vastus (13) and sartorius (1). Content leading canal (19) is rendered, not crushed.

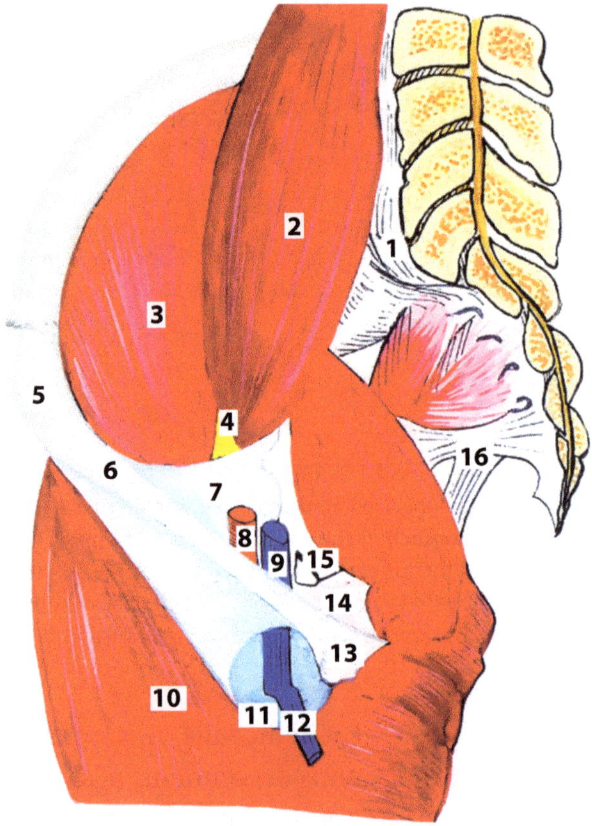

Figure 108 The femoral canal.
1 - promontorium; 2 – m. psoas major; 3 – m. iliacus; 4 – n. femoralis; 5 – spina iliaca anterior superior; 6 – lig. inguinale (Pouparti); 7 – arcus ileopectineus; 8 – a. femoralis; 9 – v. femoralis; 10 – m. sartorius; 11 – margo falciformis (cornu inferius); 12 – v. saphena magna; 13 – tuberculum pubicum; 14 – lig. lacunare; 15 – annulus femoralis internus; 16 – lig. sacrospinale et sacrotuberale

Surgical anatomy of congenital malformations. (Figure 114) Hip dislocation – broken hip ratio: the femoral head is out of the acetabulum at its first level, age, slipping on the wing of the ilium, is gradually shifting upwards and backwards. Joint capsule takes the shape of an hourglass, which creates obstacles to the reduction of dislocation. Gluteal muscle because of the displacement of the femoral head up is relaxed, gradually comes their atrophy. A child with bilateral

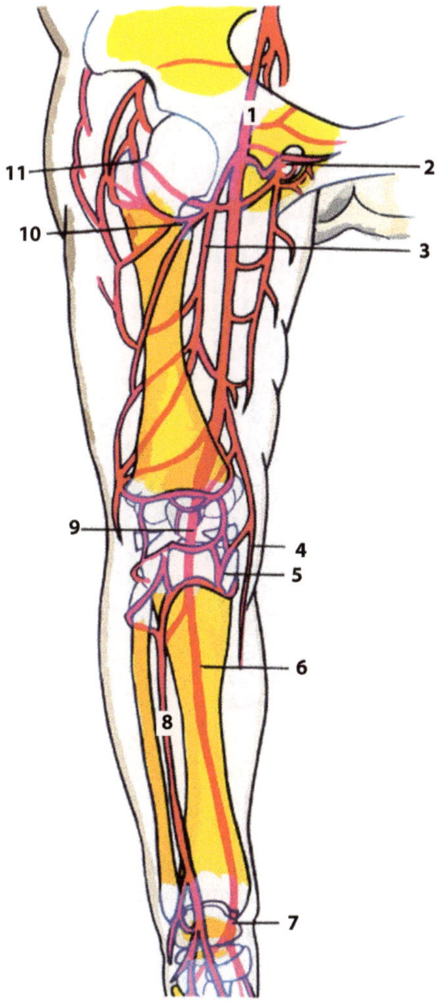

Figure 109 The lower extremity arteries. Front view.
1 – promontorium; 2 – m. psoas major; 3 – m. iliacus; 4 – n. femoralis; 5 – spina iliaca anterior superior; 6 – lig. inguinale (Pouparti); 7 – arcus ileopectineus; 8 – a. femoralis; 9 – v. femoralis; 10 – m. sartorius; 11 – margo falciformis (cornu inferius); 12 – v. saphena magna; 13 – tuberculum pubicum; 14 – lig. lacunare; 15 – annulus femoralis internus; 16 – lig. sacrospinale et sacrotuberale

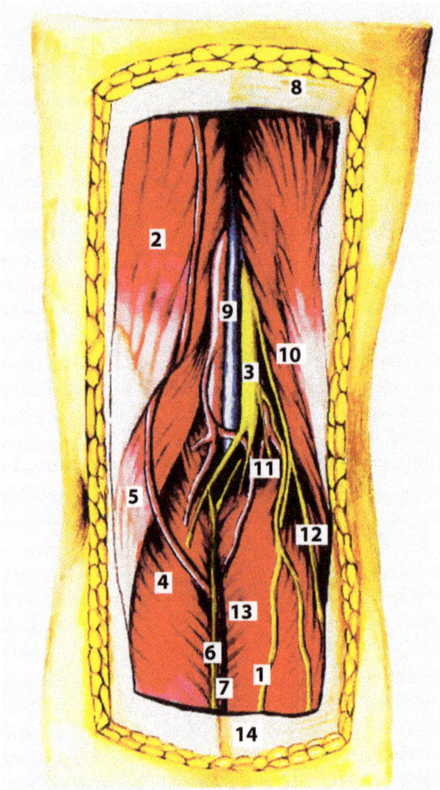

Figure 110 Topography of the popliteal fossa.
1 – anastomosis between the large and the small saphenous veins; 2 – m. semitendinosus; 3 – n. tibialis; 4 – m.gasrocnemius caput mediate; 5 – m. semimembranosus; 6 – n.gasrocnemia interna; 7 – v. saphena parva; 8 – fascia lata; 9 – a. et v. popliteal; 10 – m. biceps femoris; 11 – arterial branches to the gastrocnemius muscle; 12 – n. peroneus communis; 13 – m.gasrocnemius caput laterale; 14 – fascia cruris

congenital hip dislocation walks, swaying from foot to foot, with a unilateral dislocation limping.

Shin

Borders: upper – line drawn through tuberosita; lower – line at the base of the ankles. External signs: the front edge of the tibia, fibula head.

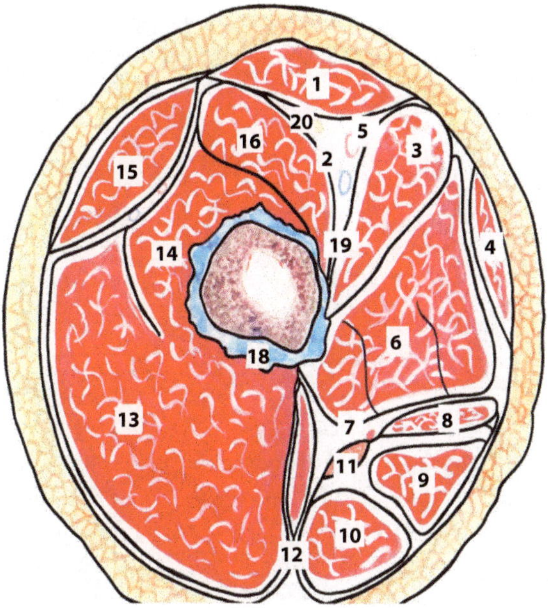

Figure 111 Cross section of the thigh. The cartilaginous tumor, cancer «*in situ*»
1 – m. sartorius; 2 – the internal intermuscular septum and the femoral vessels; 3 – long adductor muscle; 4 – tender muscle 5 – perforating artery; 6 – magnus adductor muscle; 7 – Posterior intermuscular septum; 8 – semimembranous muscle; 9 – m. semitendinosus; 10 – m. biceps; 11 – sciatic nerve; 12 – the external intermuscular septum; 13 – external vastus; 14 – medius vastus; 15 – m. rectus femoris; 16 – internal vastus; 17 – cartilaginous tumor; cancer «in situ»; 18 – the resulting channel; 19 – the saphenous nerve

There are three groups of muscles on the lower limb. Front - extensors (of feet), back – flexors, so-called peritoneal or peroneal group strengthens the arch.

The front group includes three muscles located inside from the crista tibiae outward in this order (Figure 115): m. tibialis anterior – anterior tibial muscle; m. extensor hallucis longus – long extensor digitorum longus; m. extensor digitorum longus – extensor digitorum longus.

Back muscle group is divided into two groups – superficial and deep (Figure 116). There is one big triceps muscle of shin by the surface (m. triceps surae); consisting of two heads – calf (m. gastrocnemius s. m. soleus).

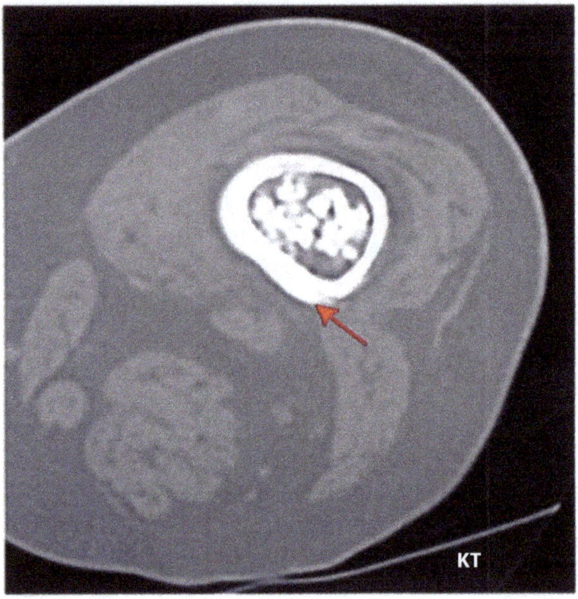

Figure 112 Cross section of the thigh. The cartilaginous tumor, cancer «*in situ*» (CT).

Deep front group is represented by antagonists of extensor: m. tibialis posterior – tibialis posterior muscle, lies in the middle; m. flexor digitorum longus – flexor digitorum longus, located medial to the previous muscle; m. flexor hallucis longus – flexor hallucis longus, located to the lateral sides of the same muscles. External group represented by two muscles – peroneus longus (m. peroneus longus), and peroneus brevis (m. peroneus brevis). Both reinforce the arch of the foot.

The blood supply of the shin. There are two main arteries: the anterior tibial artery, a. tibialis anterior, and posterior tibial artery, a. tibialis posterior. When these vessels wounded, they have to be tied up, and therefore it is necessary to know their position to be able to find them. Anterior tibial artery may be ligated in the upper, middle and lower third of the leg. Above it lies between m. tibialis anterior and m. extensor digitorum longus; in the middle and lower third of the leg, it lies between m. tibialis anterior and m. extensor hallucis longus. Throughout it is accompanied by the deep peroneal nerve (n. peroneus profundus).

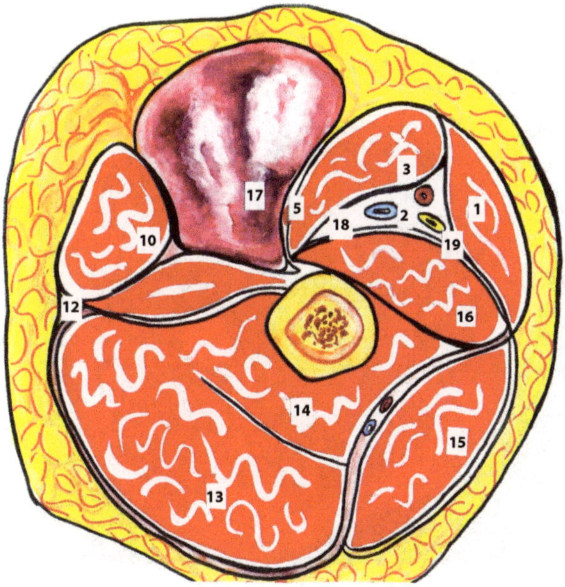

Figure 113 Cross section of the thigh. Mesenchymal fibrosarcoma of thigh.
1 – m. sartorius; 2 – the internal intermuscular septum and the femoral vessels; 3 – adductor longus muscle; 4 – tender muscle; 5 – perforating artery; 6 – adductor magnus muscle; 7 – posterior intermuscular septum; 8 – semimembranous muscle; 9 – m. semitendinosus; 10 – m. biceps; 11 – sciatic nerve; 12 – the external intermuscular septum; 13 – external vastus; 14 – medius vastus; 15 – rectus femoris; 16 – internal vastus; 17 – mesenchymal fibrosarcoma of thigh; 18 – the resulting channel; 19 – the saphenous nerve

Posterior tibial artery is located between the deep and superficial flexors, accompanied by the tibial nerve n. tibialis in all her way. On the medial surface of the tibia under the skin lies a great saphenous vein (v. saphena magna), and on the posterolateral side of the shin – small subcutaneous vein (v. saphena parva). These veins are often subjected to varicose and require special surgery treatment.

Zone of sensory and motor innervation of the n. peroneus superficialis. The nerve descends into the upper canal of the peroneal muscle – up to mid-calf level, pierces the fascia of the leg and goes into the body of fat tissue. It innervates the lateral group of leg muscles and skin of the lateral surface of the leg and the posterior foot. Zone of sensory and motor innervation of the n. peroneus profundus. The nerve goes

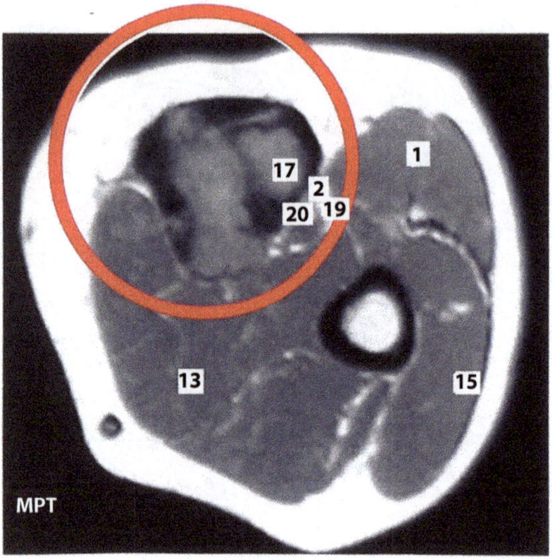

Figure 114 Cross section of the thigh. Mesenchymal fibrosarcoma of thigh. (MRI.).

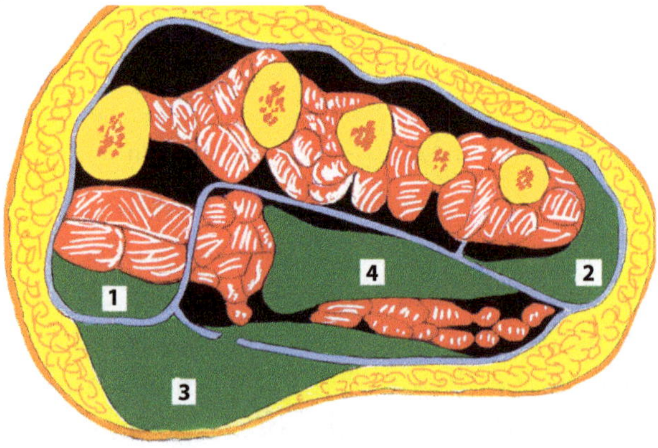

Figure 115 The relationship of the femoral head and the acetabulum of the hip capsule. Frontal section of limb in flexion position (by DD Kupatadze).
1 - acetabulum; 2 – capsula arlicularis; 3 – caput femoris; 4 – lig. capitis femoris

around the neck of the fibula, anterior intermuscular septum pierces and lies on the interosseous membrane in the upper third of the leg between the tibialis anterior and extensor digitorum longus, lateral to the anterior tibial artery. Innervates the anterior group of leg muscles

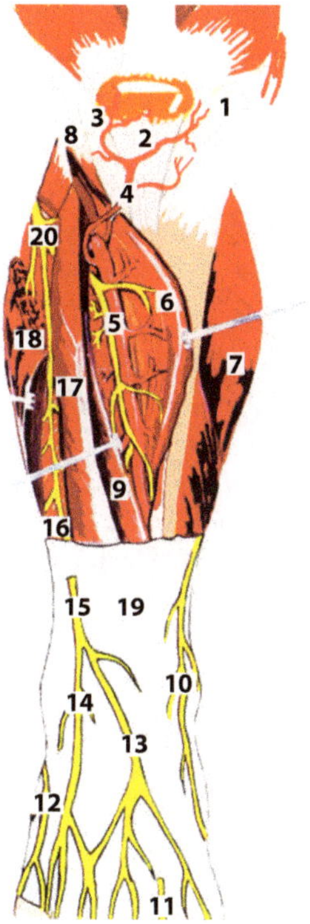

Figure 116 The front surface of the shin.
1 – retinaculum patellae mediaie; 2 – lig. patellae; 3 – retinaculum patellae laterale; 4 – a. recurrens tibialis anterior; 5 – a. tibialis anterior et n. peroneus tibularis profundus; 6 – m. tibialis anterior; 7 – m. gasrocnemius; 8 – caput tibilae; 9 – m. extensor hallucis longus; 10 – n. cutaneus internus; 11 – ramus cutaneus nervi fibularii profundi; 12 – rami terminalis n. surali; 13 – ramus medianus cutaneus dorsalis pedis; 14 – ramus intermedialis cutaneus dorsalis pedis; 15 – ramus superficialis n. fibularis; 16 – m. peroneus brevis; 17 – m. extensor digitorum longus; 18 – m. peroneus longus; 19 – fascia cruris; 20 – n. fibularis communis

and skin of the first interdigital spaces on foot. N. tibialis innervates all muscles of the back group.

Cruropoplit canal (canalis cruropopliteus). The front septum – tibialis posterior muscle, flexor digitorum longus. The posterior septum – sheet of deep fascia and the soleus muscle adjacent to it. A hole in the top canal includes a. et v. poplitea and n. tibialis (Figures 118). Anterior tibial artery and vein pass through the front opening, enter

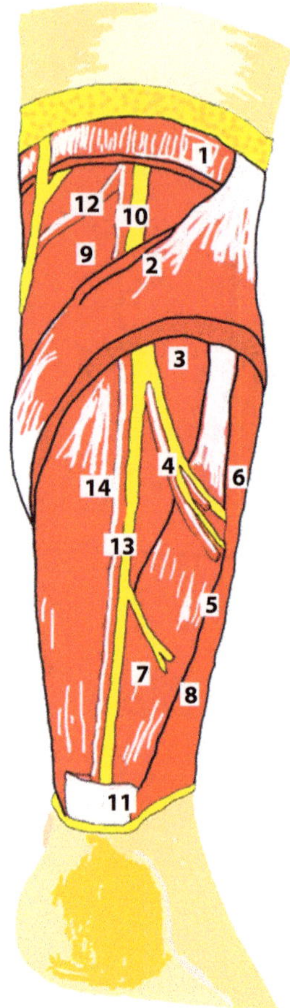

Figure 117 The posterior area of the shin.
1 – m. gastrocnemius; 2 – m. soleus et arcus tendineus; 3 – m. tibialis posterior; 4 – a. fibularis; 5 – septum intermuscularis posterior; 6 – m. peroneus longus; 7 – m. extensor hallucis longus; 8 – m. peroneus brevis; 9 – m. popliteus; 10 – a. poplitea et n. tibialis; 11 – tendo Achillis; 12 – a. genus inferior medialis; 13 – a. tibialis posterior et n. tibialis; 14 – m. flexor digitorum longus

the front muscle bed. After a lower opening passes posterior tibial artery and vein goes n. tibialis. (Figures 112 and 113).

Canals of shin. Supraperoneal muscular canal (canalis musculoperoneus superior) is between the fibula and peroneus longus. There is n. peroneus communis, which is divided by n. peroneus superficialis et profundus. Infraperoneal muscular canal (canalis musculoperonaeus inferior) laterally limited fibula, in the back – the long flexor of the

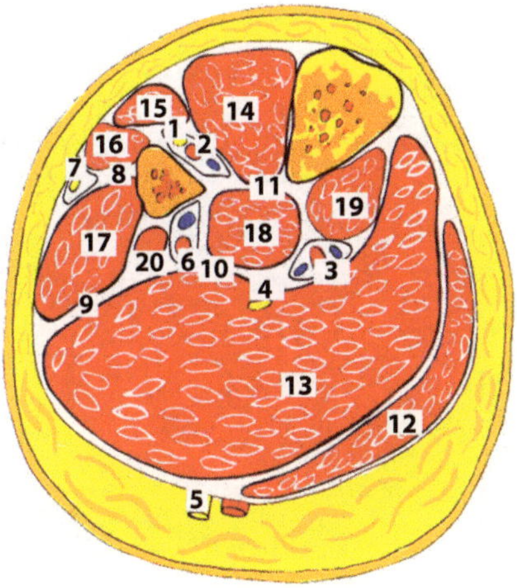

Figure 118 The cross-cut of the upper third of the shin.
1 – n. peroneus profundus; 2 – a. tibialis anterior; 3 – a. tibialis posterior; 4 – n. tibialis; 5 – n. cutaneus surae internus; 6 – a. fibularis; 7 – n. fibularis superllcialis; 8 – septum intermuscularis anterior; 9 – septum intermuscularis posterior; 10 – lamina fasciae cruris profunda; 11 – membrana interossea; 12 – m. gastrocnemius; 13 – m. soleus; 14 – m. tibialis anterior; 15 – m. extensor hallucis longus; 16 – m. extensor digitorum longus; 17 – mm. fibulares; 18 – m. tibialis posterior; 19 – m. flexor digitorum longus; 20 – m. flexor hallucis longus

toe, front – m. tibialis posterior. The canal includes. a et v. m. tibiales posterior (Figures 121 and 122 on the MR-visible knee-popliteal canal of tibia (22), comprising a posterior tibial artery and vein (6) and the tibial nerve (22)).

Neurovascular bundle. Neurovascular bundle of the popliteal fossa. As part of the neurovascular bundle – a. et v. poplitea, n. tibialis. Front and medially located artery, lateral to and behind it – vein, more lateral and back – nerve (the mnemonic code NeVA): a. poplitea enters the popliteal fossa of the lower opening of the leading canal, then enters the upper foramen canalis crurapopliteus; v. poplitea lies in the popliteal fossa lateral and behind the artery of the same name; n. tibialis located in the popliteal fossa and behind the lateral popliteal vein, branches depart from it to the back of the lower leg muscle group and n. cutaneus surae medialis. n. peroneus communis extends from the lateral border of the popliteal fossa.

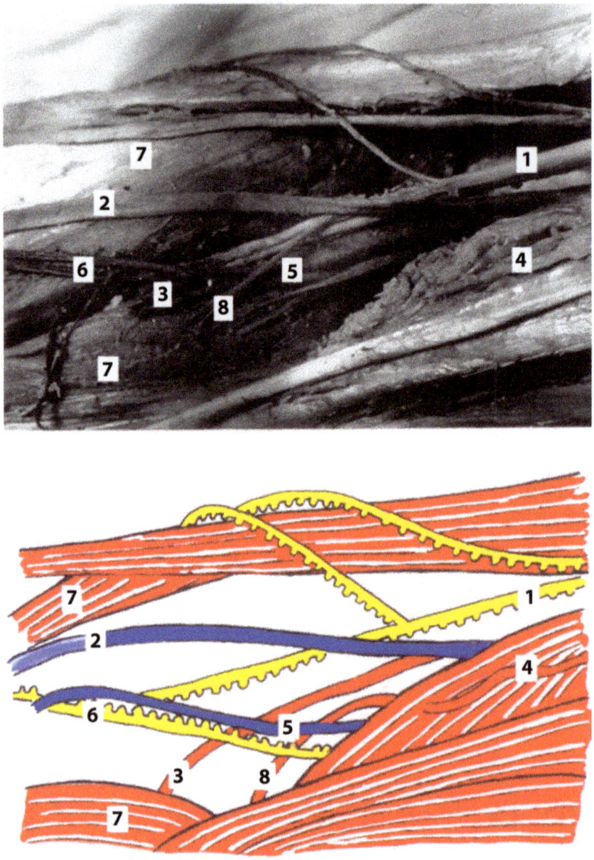

Figure 119 Topography of the upper third of the back of the shin.
1 – n. tibialis; 2 – v. saphena parva; 3 – a. tibialis posterior; 4 – a. recurrens tibialis posterior; 5 – v. tibialis posterior; 6 – n. cutaneus surae medialis; 7 – m. soleus; 8 – a. peronea

Knee joint (articulatio genus) (Figures 123–125) is located form the distal epiphysis of the femur, proximal epiphysis of the tibia and the patella. The distal end of the femur is extended downward to form condylus medialis et condylus lateralis, for articulation with the tibial plateau. The articular capsule consists of fibrous and synovial membranes:

1. membrana fibrosa strengthens ligaments: lig. patellae and retinaculum patellae mediale et retinaculum patellae

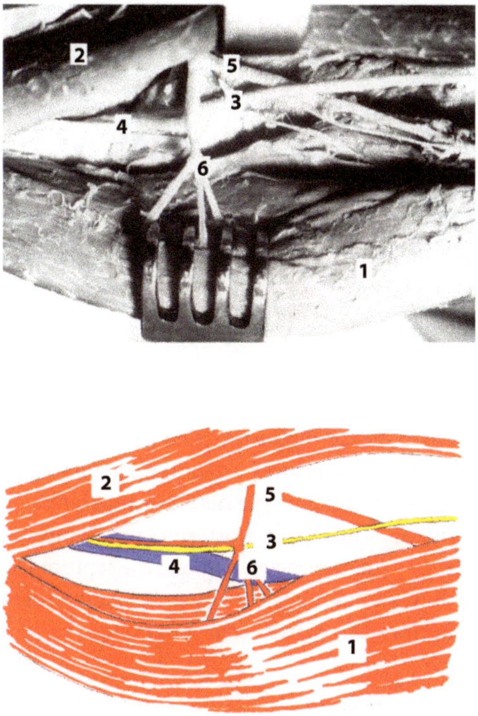

Figure 120 Topography - vascular nerve bundle upper third of the back of the leg. 1 – m. soleus; 2 – m. tibialis posterior; 3 – n. peroneus; 4 – v. tibialis posterior; 5 – a. tibialis posterior; 6 – a. fibularis

laterale, lig. collaterale fibulare, lig. popliteum obliquum, lig. popliteum arcuatum, lig. transversum genus).

2. membrana synovialis forms joint folds inwards: plica synovialis infrapatellaris, plicae alares, bursa infrapatellaris profunda. Intracapsular ligament: lig. cruciatum anterius, lig. cruciatum posterius, lig. meniscofemorale posterius.

Bursas: bursa suprapatellaris, bursa anserina, bursa subtendinea m. gastrocnemii medialis, bursa subtendinea m. lateralis.

The innervation of the joint takes place on n. peroneus – external, n.femoralis – internal, n. tibialis – behind.

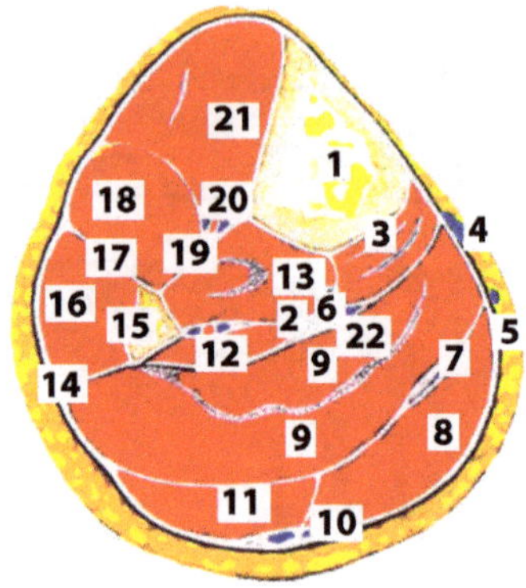

Figure 121 A cross section of the shin.
1 – os tibia; 2 – n. tibialis; 3 – m. flexor digitorum longus; 4 – v. saphena magna; 5 – fascia cruris; 6 – a. et v. tibialis posterior; 7 – tendo m. plantaris; 8 – caput mediale m. gastrocnemii; 9 – m. soleus; 10 – v. saphena parva; 11 – caput laterale m. gastrocnemii; 12 – m. flexor hallucis longus; 13 – m. tibialis posterior; 14 – septum intermusculare cruris posterius; 15 – fibula; 16 – m. peroneus longus; 17 – septum intermusculare cruris anterius; 18 – m. extensor digitorum longus; 19 – membrana interossea; 20 – a. et v. tibialis anterior; 21 – m. tibialis anterior; 22 – canalis cruropopliteus

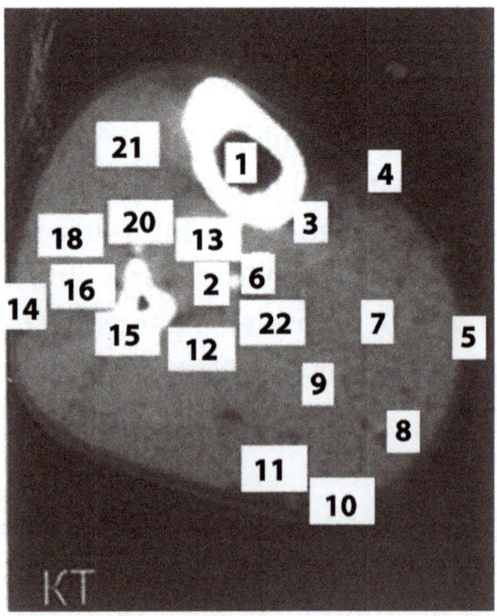

Figure 122 A cross section of the shin (CT).

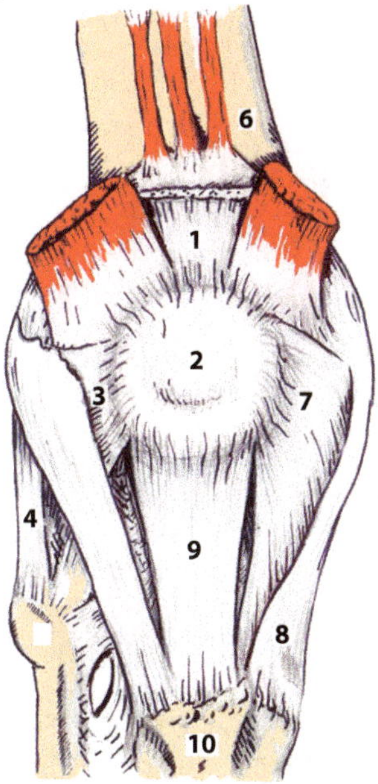

Figure 123 The knee joint. Front view.
1 – tendo m. rectus femoris; 2 – patella; 3 – retinaculum patellae laterale; 4 – lig. collaterale fibulare; 5 – fibula; 6 – os femoris; 7 – retinaculum patellae mediale; 8 – lig. collaterale tibiale; 9 – lig. patellae; 10 – tibia

Blood supply of the knee is performed by rete articularis genus (Figure 126).

Lymph vessels goes from the front of the knee occur in the inguinal lymph nodes. On the back – in the popliteal lymph nodes.

Features of the structure of the knee in children: The joint of a newborn has a rounded shape, the contours of the patellar ligament and unobtrusive. Epiphysis of the femur and tibia forming the knee joint, the cartilage tissue formed containing nuclei of ossification; patella fully formed cartilage. The outer femoral condyle is developed more than the inner. By the age of two sizes of condyles are aligned. Joint capsule strengthens the front patellar tendon. Bias and the arcuate popliteal ligament weak. Cruciate ligament thick, short.

Surgical anatomy of congenital malformations. Clubfoot – congenital contractures of the joints of the foot. The foot is in flexion in the

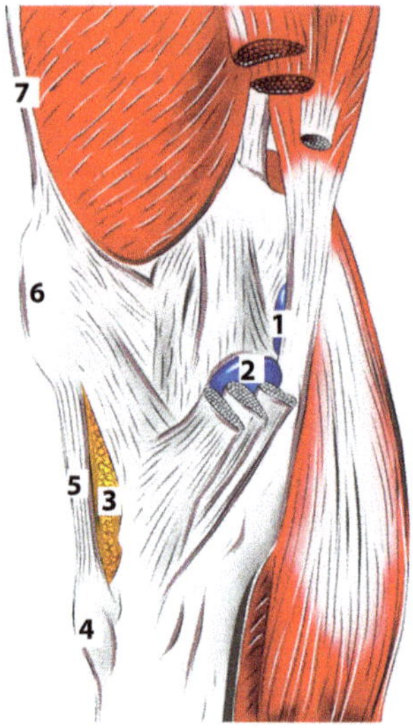

Figure 124 The knee joint. The view from the medial side.
1 – bursa m. semimembranosi; 2 – bursa anserina; 3 – corpus adiposum infrapatellare; 4 – tibia; 5 – lig. patellare; 6 – patella; 7 – tendo m. quadricipitis femoris

ankle joint, its outer edge is lowered, and the front office is given and bent with increasing longitudinal arch. Ligaments and joint capsule on the medial side of the foot atrophy. Tibialis posterior muscle and flexor hallucis longus are shortened, and their tendons are thickened. Heel tendon is also thickened. Peroneus longus usually hypertrophic, their tendons are shifted backward.

Foot

Extensor muscles are located on the posterior of the foot (dorsum pedis) (Figure 127); a powerful development flexor system lies on the sole of the foot, planta pedis. Figure 121 shows a sagittal section of the foot. The all anatomical structures identified in the figure caption at MR tomogram (Figure 127). We can see the following bones relationship: the shank (9), followed by the tarsal bones – the lateral cuneiform bone (8), the navicular (6), the intermediate sphenoid bone (7), cuboid (5), astragalus (3), calcaneus (4). On the next MRI CT scan of the patient are visualized heel spurs (16) unlike the normal slice, the

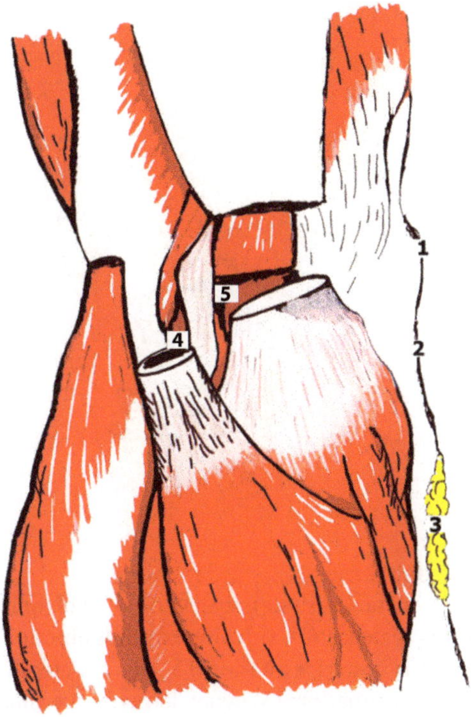

Figure 125 The knee joint. View from the lateral side.
1 – patella; 2 – lig. patellae; 3 – corpus adiposum infrapatellare; 4 – bursa subtendinea m. bicipitis femoris inferior; 5 – bursa subtendinea

change is in shape of the contour of the calcaneus (4) with compression of the underlying muscles, tapping the little finger (10). There are special canals formed behind the ankles during the transition to the foot that enclose the tendons or blood vessels and nerves. Thus, there are four canals over the medial malleolus. In the first, the second and fourth are located the flexor tendons, in the third – the neurovascular bundle: tibial vessels, vasa tibialia posteriora (two veins and arteries) and the tibial nerve, n. tibialis. Near the lateral malleolus lies two peroneal tendons muscles. The boundaries: a horizontal line drawn through the base of the ankles. On the foot allocated regio dorsalis pedis and regio plantaris pedis. External guidance: malleolus lateralis et medialis, tuber calcaneus, I and V metatarsal bones.

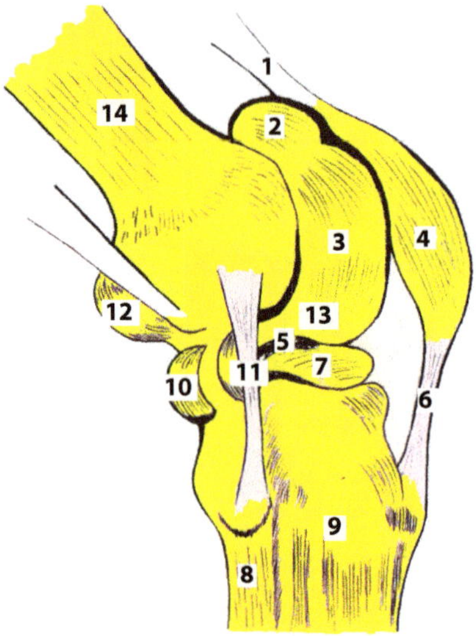

Figure 126 Volvulus of the right knee.
1 – tendo m. quadriceps; 2 – bursa serosa suprapatellaris; 3 – volvulus superius; 4 – patella; 5 – meniscus externus; 6 – lig. patellae; 7 – volvulus externus anterior inferior; 8 – fibula; 9 – tibia; 10 – volvulus externus posterior inferior; 11 – lig. collaterale externum; 12 – volvulus externus posterior superior; 13 – volvulus externus anterior inferior; 14.femur

Neurovascular bundle (Figure 130). **The projection on the skin surface, syntopy their elements**. The posterior part of the foot: here are a. et v. dorsales pedis and n. peroneus profundus. The artery lies between the tendons mm. extensor digitorum longus et extensor hallucis longus. Before reaching the first interdigital spaces a. dorsales pedis gives a. arcuata, which runs under the m. extensor hallucis brevis and gives aa. metatarseae dorsales, which are formed from aa. digitales dorsales, a. metatarseae dorsalis prima and ramus plantaris profundus, which goes on the foot through the first period intertarsal and participates in the formation of arcus plantaris. The nerve is located medially from the arteries and gives a branch to related parties 1 and 2 fingers. It innervate the extensor digitorum brevis.

The sole of the foot (Figure 131): the medial neurovascular bundle (a., v., n. Plantares mediales) goes to the sulcus plantaris medialis. The lateral neurovascular bundle (a., v., n. plantares laterales) – goes into the sulcus plantaris lateralis. The arcus plantae enters at the level of the

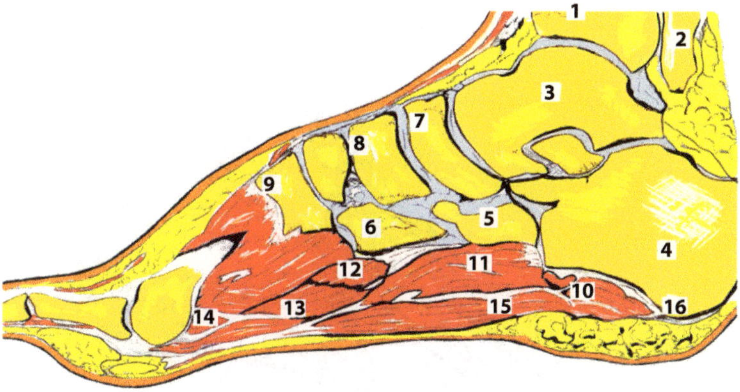

Figure 127 Sagittal section of the foot.
1 – os tibia; 2 – os fibula; 3 – talus; 4 – calcaneus; 5 – os cuboideum; 6 – os naviculare; 7 – os cuneiforme intermedium; 8 – os naviculare laterale; 9 – metatarsus; 10 – m. abductor digiti minimi; 11 – m. flexor digitorum brevis; 12 – m. adductor hallucis (caput obliquum); 13 – m. lumbricalae; 14 – m. adductor hallucis (caput transversum); 15 – n.; a. et v. plantaris lateralis; 16 – the heel spur

base of the metatarsal artery. Anastomosis doug with the deep plantar branch of a. dorsalis pedis. They are moving away aa. metatarsea et plantares, which formed from aa. digitales plantares. Plantar region is innervated by n. plantaris medialis et lateralis.

Blood supply of the posterior of the foot is realized at the expense of dorsal artery of foot, a. dorsalis pedis (Figures 132 and 133). It should be first in the direction of the interdigital lacunas.

Posterior tibial artery, a. tibialis posterior, larger cross-section; rounded the medial malleolus. It is divided into the main terminal branches – the internal plantar artery, a. plantaris medialis, and an outer plantar artery, a. plantaris lateralis. If they are damaged, they produced a ligation of the posterior tibial artery.

Ankle (articulatio talocruralis) (Figures 134 and 135) is formed by the distal epiphyses of shin bones and the talus, which are interconnected by anterior and posterior tibiofibular ligaments. The shape of the joint is ginglymoid. The articular capsule attached on the edge of the articular cartilage, ligaments are reinforced laterally. Lig. mediale moves on the medial malleolus. There are three ligaments depart from the lateral ankle: lig. talofibulare anterius, lig. talofibulare posterius,

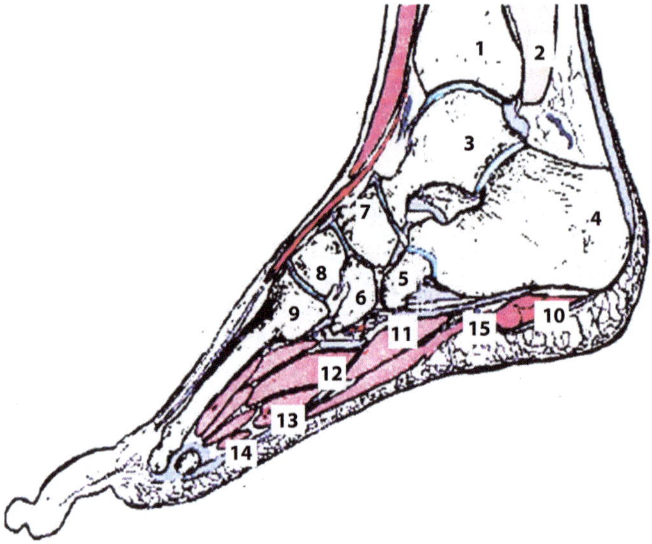

Figure 128 Sagittal section of the foot.
1 – os tibia; 2 – os fibula; 3 – talus; 4. calcaneus; 5 – os cuboideum; 6 – os naviculare; 7 – os cuneiforme intermedium; 8 – os naviculare laterale; 9 – metatarsus; 10 – m. abductor digiti minimi; 11 – m. flexor digitorum brevis; 12 – m. adductor hallucis (caput obliquum); 13 – m. lumbricalae; 14 – m. adductor hallucis (caput transversum); 15 – n.; a. et v. plantaris 16 – the heel spur

lig. calcaneofibulare. The blood supply of the anterior joint capsule is carried by a. tibialis anterior, posterior – a. tibialis posterior, and exterior by a. peronea. N. saphenus, n. suralis, and n. peroneus profundus are involved in the innervation of the joint capsule.

The topography of the saphenous veins and nerves of the lower limbs is visualized in Figures 136–138.

Canals of foot. The medial ankle channel (canalis malleolaris medialis) has four abodes of fibrotic: tendo m. tibialis posterior passes through the first channel behind the ankle; through the second – tendo m. flexoris digitorum longi; through the third pass a. et v. posteriores tibialis and n. tibialis; through the fourth channel – tendo m. flexor hallucis longus.

Lateral ankle channel (canalis malleolaris lateralis) serves as a receptacle for tendo mm. peronaei longus et brevis.

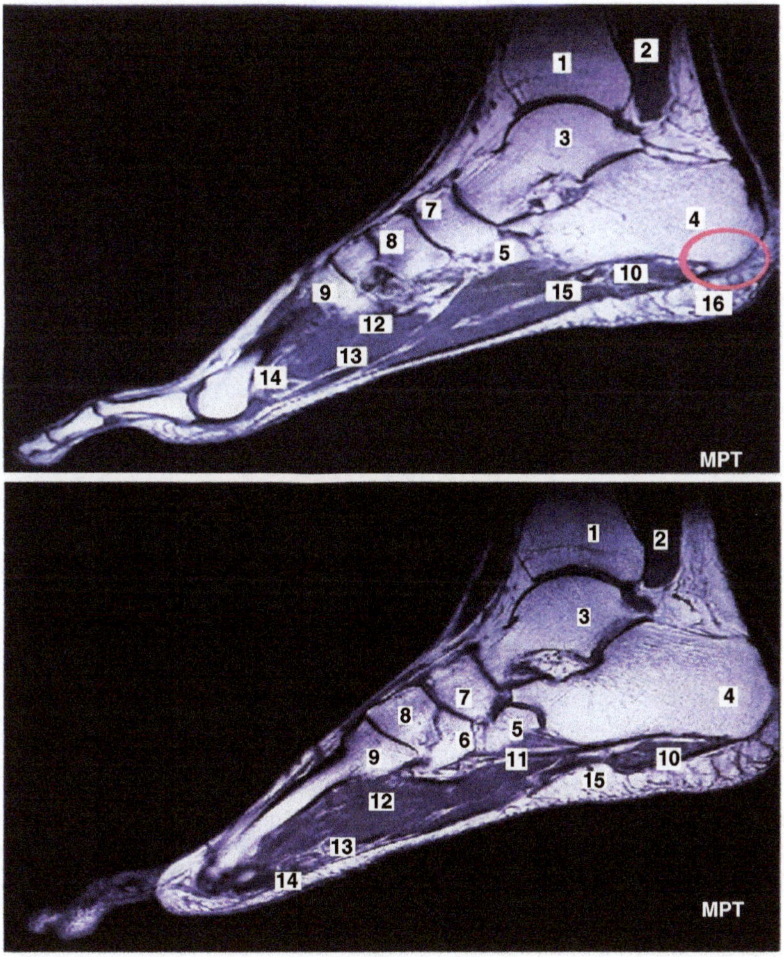

Figure 129 Sagittal section of the foot (MRI).

Ankle posrterior channels (canalis talocruralis). tendon m. tibialis anterior goes through the medial canal; m. extensor hallucis longus goes through the middle canal, a. et v. tibialis anterior, n. peroneus profundus; 4 tendons of m. extensor hallucis longus pass through the lateral canal.

Features of the structure of the children's joint: neonatal joint is formed by cartilaginous epiphysis shin bone and cartilage of the talus. The capsule of the ankle joint is very thin, the ligaments are poorly developed, especially the deltoid.

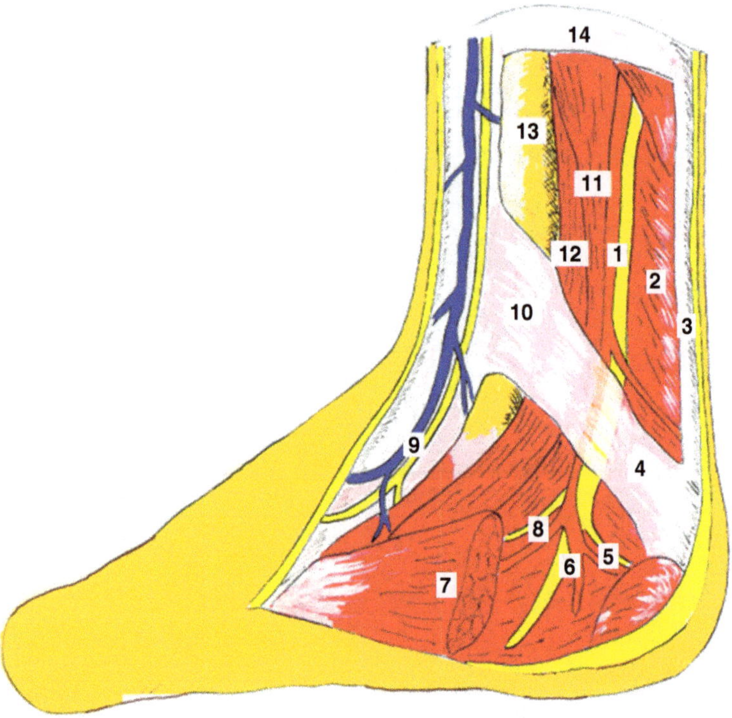

Figure 130 Area of the inner ankle.
1 – a. et n. tibialis; 2 – m. flexor hallucis longus; 3 – tendo Achillis; 4 – lig. retinaculi musculorum flexorum; 5 – ramus calcanei; 6 – n. et a. plantaris extensor; 7 – m. abductor hallucis; 8 – n. et a. plantaris internus; 9 – v. et n. saphena magna; 10 – malleolus interims; 11 – m. flexor digitorum longus; 12 – m. tibialis posterior; 13 – os tibialis; 14 – fascia cruri

Heel channel is the space between the heel bone and the abductor muscle of the thumb. The dense aponeurotic partition divides it into two sections: anterior and posterior. M. quadratus plantae, the external plantar neurovascular bundle passing in splitting the fascia are located in the back. Tendon flexor hallucis longus and flexor digitorum longus, in the partition that separates the square from the muscle tendon passes medial plantar neurovascular bundle are located in the front.

Plantar channel (canalis plantaris) is located in the proximal deep fascial space mid-sole bed. Borders: upper – lig. plantare longum, lower – deep fascia covering the flexor digitorum longus muscle to the square, the lateral septum of the outer box and the m. quadratus

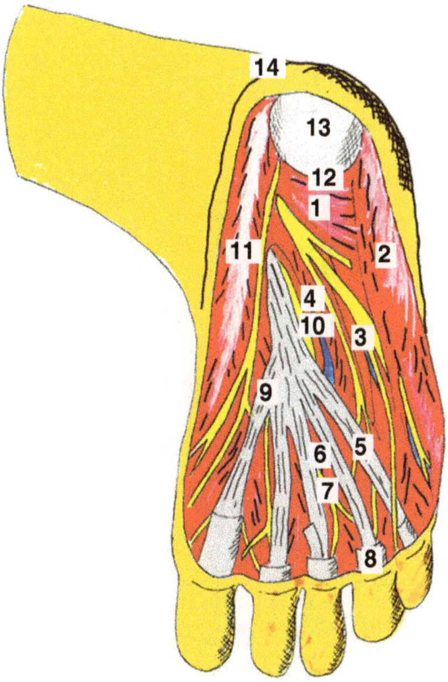

Figure 131 The sole of the foot.
1 – m. flexor digitorum brevis; 2 – m. abductor digiti minimi; 3 – n. digitalis plantaris communis; 4 – m. qudratus plantae; 5 – m. lumbricalis; 6 – m. flexor digitorun longus; 7 – vasis. et n. plantaris externi; 8 – tendo m. flexor digitorum brevis; 9 – tendo m. flexor digitorum longus; 10 – n. et vasis plantaris internus; 11 – m. abductor hallucis; 12 – aponeurosis plantaris; 13 – tuber calcanei; 14 – tendo Achilis

plantae, medial – inner septum of the bed. In the channel lies loose fiber, square muscle and plantar, lateral and medial neurovascular bundles.

Pathotopography of the Lower Limbs

Mesenchymal fibrosarcoma of the thigh is an immature, malignant fibroblastic neoplasm of connective tissue origin. It is a painless, round, small-hummocky knot. It consists of immature fibroblast-like cells and collagen fibers. The source of mesenchymal tumors is the polypotent mesenchymal cell. Depending on the degree of maturity and the relationship between the cellular and fibrous elements of the tumour, differentiated and low-grade fibrosarcoma are distinguished.

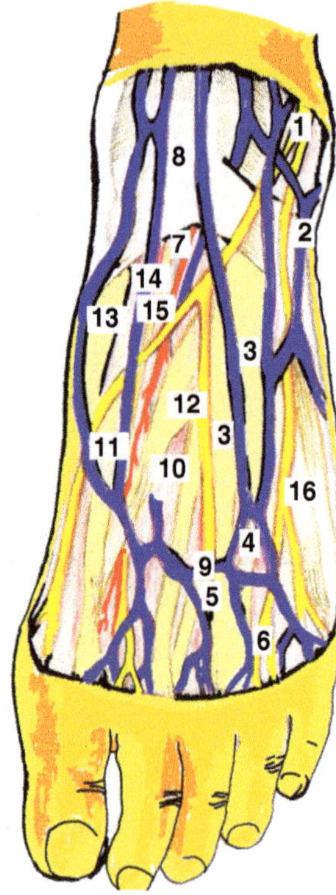

Figure 132 Surface vessels and nerves of the ankle and posterior foot.
1 – n. peroneus superficialis; 2 – v. saphena parva; 3 – tendines mm. extensoris digitorum longi; 4 – tendines mm. extensoris digitorum brevis; 5 – n. cutaneus dorsalis intermedius; 6 – n. cutaneus dorsalis lateralis; 7 – a. dorsalis pedis; 8 – retinaculum mm. extensorum inferius; 9 – arcus venosus dorsalis pedis; 10 – mm. interossei dorsales; 11 – m. extensor hallucis longus; 12 – m. extensor hallucis brevis; 13 – tendo m. tibialis anterioris; 14 – v. saphena magna; 15 – n. peronaeus profundus; 16 – tendo m. peroneus brevis

The tumor node (low-diffused fibrosarcoma) does not have clear boundaries (infiltrating growth), soft consistency, on the cut its gray-pink color, resembles "fish meat", with foci of necrosis and hemorrhages. Differential fibrosarcoma sometimes has a false capsule, which is a pushed and squashed local stroma. Fibrosarcoma, in comparison with other sarcomas, metastasizes less often, but recurs more often than other tumors (metastasizes primarily hematogenous to the lungs, less often to the liver, then lymphogenically to regional lymph nodes).

Pathotopography (Figures 139 and 140). It develops in the thickness of muscles, can be asymptomatic for a long time. Skin over the tumor

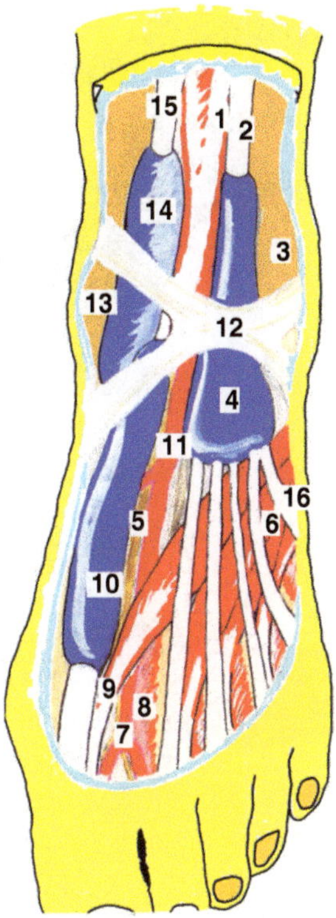

Figure 133 Anterior area of the ankle joint.
1 – tendo m. extensor hallucis longus; 2 – tendo m. extensor digitorum longus communis; 3 – malleolus externus; 4 – vag. synovialis tendinis m. extensor digitorum communis; 5 – n. tibularis profundus; 6 – mm. interossea; 7 – a. metatarseae prima; 8 – m. extensor ditorun brevis communis; 9 – tendo m. extensor hallicis brevis; 10 – vag. synovialis tendinis; 11 – a. dorsalis pedis; 12 – lig. retinaculi musculori extesiri; 13 – maleollus externus; 14 – vag. synovialis m. tibiali anterior; 15 – tendo m. tibialii anterior; 16 – tendo m. fibularii tetrii

is not changed. Reaching considerable size, causes deformation of the affected area. With fast-growing large and superficially located fibrosarcomas, thinning of the skin, a cyanotic skin tone and an extended network of subcutaneous veins in the neoplasm can be marked. A characteristic feature is the impression of a limited node, "false capsule" or pseudocapsule, which is a layer of fibrous tissue. With progression, the false boundaries of the tumor become less pronounced.

Small local formations may shift (more often in the transverse direction). When the surrounding tissues grow, the fibrosarcoma becomes

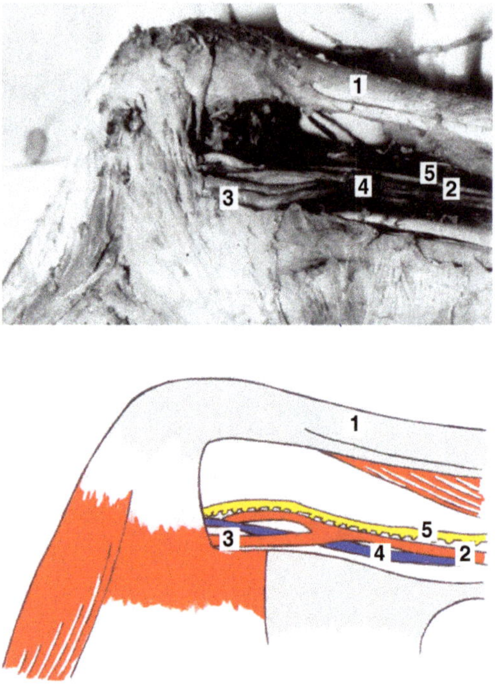

Figure 134 Topography of heel.
1 – tendo calcaneus Achillis; 2 – a. tibialis posterior; 3 – a. plantaris lateralis; 4 – v. saphena magna; 5 – n. tibialis

immobile. With compression of the nerves, tenderness is noted during palpation. When the bone is damaged, the pain becomes permanent. In the later stages of fibrosarcoma, symptoms of general intoxication are revealed. The patient loses weight and appetite. There is a rise in temperature, anemia, increasing weakness, emotional lability, depression or subdepression.

The degree of mobility of fibrosarcoma is determined by the prevalence of the process. Small local formations may shift (more often in the transverse direction). When the surrounding tissues grow, the fibrosarcoma becomes immobile. When the neoplasm is located in the intermuscular space, the nodes are well palpated during muscle relaxation, lose contours and lose mobility with muscle tension. In the initial stages, fibrosarcomas are usually painless. With compression of

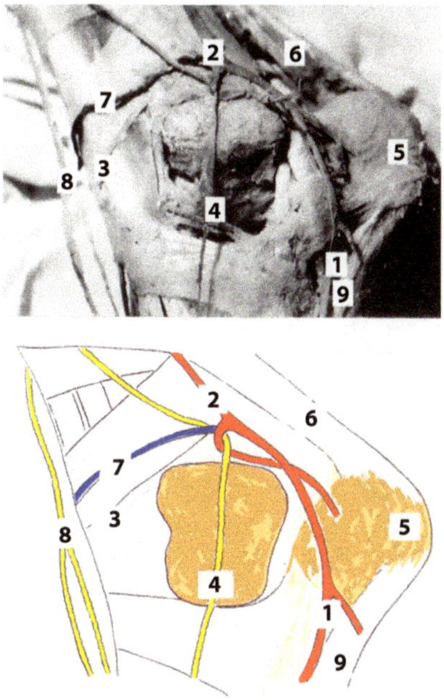

Figure 135 Topography of the medial surface of the ankle joint.
1 – a. tibialis posterior; 2 – a. plantaris medialis; 3 – retinaculum mm. extensorum superior; 4 – n. communicans; 5 – tuber calcanei; 6 - tendo calcaneus Achillis; 7 – arcus venosus dorsalis; 8 – n. cutaneus dorsalis; 9 – lig.plantaris longus

the nerves, tenderness is noted during palpation. When the bone is damaged, the pain becomes permanent.

Chondrosarcoma is a malignant tumor of a cartilaginous structure. Chondrosarcoma of peripheral localization is well revealed in the form of a dense tumor, adjacent to the cortical layer of bone, sometimes sprouting into the bone at different depths. There may be foci of ossification and petrification.

A tumor of a densely elastic consistency, in the presence of mucus, decay or hemorrhage – softer, does not have a clear restriction from neighboring tissue. It is observed mainly in middle and old age, and in men, it is twice as likely as in women. The tumor can occur in any bone (be it long tubular, flat or short), develops from the cartilage through the enchondral ossification. It is more often localized in the pelvic bones, shoulder girdle and ribs, less often in the limbs and spine. Often affects the proximal ends of the femur and humerus. In 10–15% of

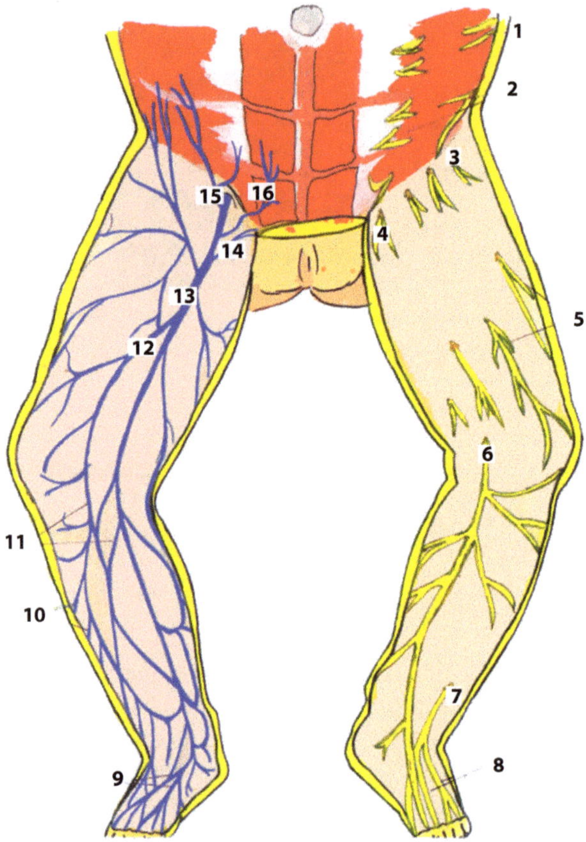

Figure 136 Subcutaneous veins and nerves of the lower extremities.
1 – n. subcostalis; 2 – n. iliohypogastricus; 3 – n. cutaneus femoris lateralis; 4 – n. genicofemoralis; 5 – rr. cutanci anteriores n. femoralis; 6 – n.saphenus; 7 – n. peroneus superficialis; 8 – n. cutaneus dorsalis intermedius; 9 – rete venosum dorsalis pedis; 10 – vv. peroneae; 11 – vv. genus; 12 – v. saphena accessoria; 13 – v. saphena magna; 14 – v. pudenda externa; 15 – v. circumflexa ilium superficialis; 16 – v. epigastrica superficialis

patients, chondrosarcomas develop secondary from previous pathological processes: osteochondromatosis, enchondromas, echondromas, solitary osteochondromas, Ollier's disease (dyskondroplasia), Paget's disease (deforming osteodystrophy). In 15% of patients, there are fast growing chondrosarcomas, prone to accelerated growth and rapid malignancy.

Pathotopography (Figures 141 and 142). Chondrosarcomas are characterized by an asymptomatic gradual course. At the initial stage

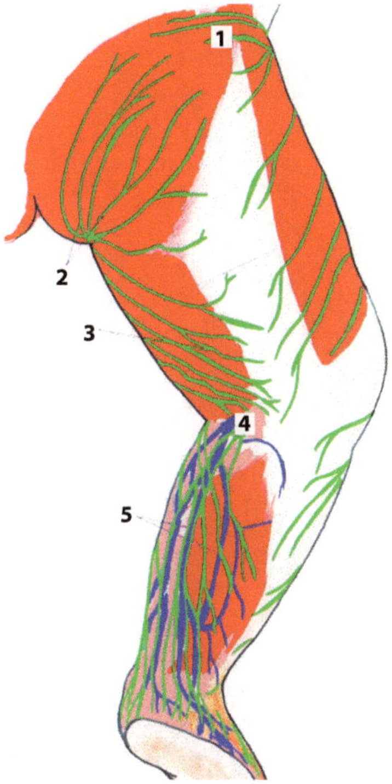

Figure 137 Superficial lymph vessels of the lower limb. Back view.
1 – vasa lymphatica iliaca; 2 – vasa lymphatica clunii inferior; 3 – vasa lymphatica superficialia femoralia; 4 – nodi lymphatici poplitales; 5 – v. saphena parva et vasa lymphatica superficialia cruris

of development pathological signs are absent. With the growth of the tumor, pains begin to occur in the localization of chondrosarcoma. If it is located near the joint, then there are signs of movement restriction in the clinic of the disease. If the tumor begins to squeeze the sciatic nerve, the pain changes from insignificant to intolerable, giving off in the buttock and on the back surface of the thigh. The tumor reaches a significant size, it can be easily felt, the skin above it becomes edematous.

Tumors of large size can grow in different directions and lead to displacement and compression of muscles: outer, middle, inner wide muscles, long and large adductor muscles, a straight muscle of the thigh, which will lead to their atrophy. Displacement of these muscles can protect the contents of the leading canal - the femoral artery and

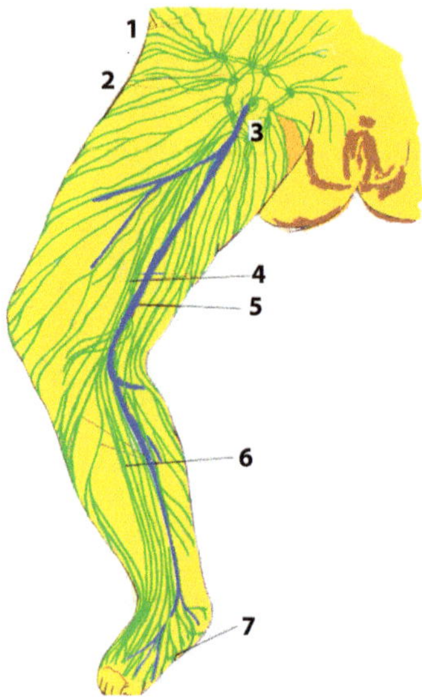

Figure 138 Superficial lymph vessels of the lower limb. Front view.
1 – vasa lymphatica circumflexa ilium; 2 – nodi lymphatici inguinales superficialis superomediales; 3 – nodi lymphatici inguinales superficialis superolaterales; 4 – vasa lymphatica femoralia superficialia; 5 – v. saphena magna; 6 – vasa lymphatica cruris superficialia; 7 – rete limphaticum plantare

vein, the perforating artery, the subcutaneous nerve, which causes characteristic pains, ischemia, edema.

In all types of large-sized chondrosarcomas located in the pelvic bones region, a number of characteristic symptoms may occur, due to tumor pressure on adjacent organs and tissues. When the sciatic plexus is compressed, pain occurs. With compression of the neck of the bladder, difficulty urinating. When the iliac vein is compressed, one-sided swelling of the lower limb can develop on the side of the lesion.

Chondrosarcoma is especially dangerous when it forms in the base of the skull, because it squeezes the brain structures when the intracranial position is located. When localized in the nasal sinus area causes difficulty in nasal breathing, with the formation of tubular bones in the epiphyses, it can cause synovitis. In some cases, pathological fractures are possible.

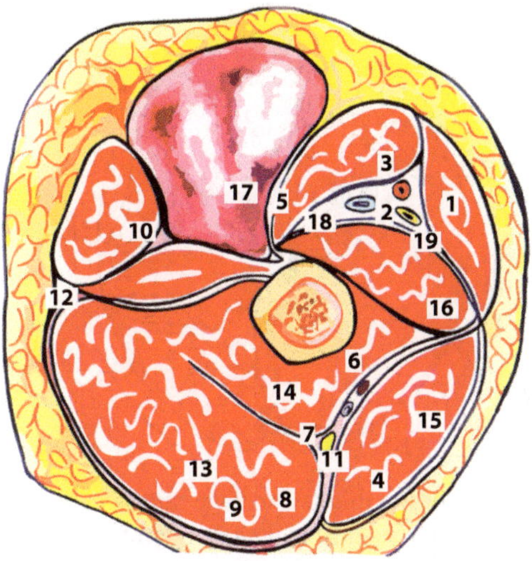

Figure 139 Transverse section of the thigh. Mesenchymal fibrosarcoma of the thigh. 1 - sartorius muscle; 2 - internal intermuscular septum and femoral vessels; 3 - long adductor muscle; 4 - tender muscle; 5 - perforating artery; 6 - large adductor muscle; 7 - posterior intermuscular septum; 8 - semimembranous muscle; 9 - semitendinous muscle; 10 - biceps muscle; 11 - sciatic nerve; 12 - external intermuscular septum; 13 - external broad muscle; 14 - medium broad muscle; 15 - straight muscle of the thigh; 16 - internal broad muscle; 17 - mesenchymal fibrosarcoma of the thigh; 18 - leading channel; 19 - subcutaneous nerve

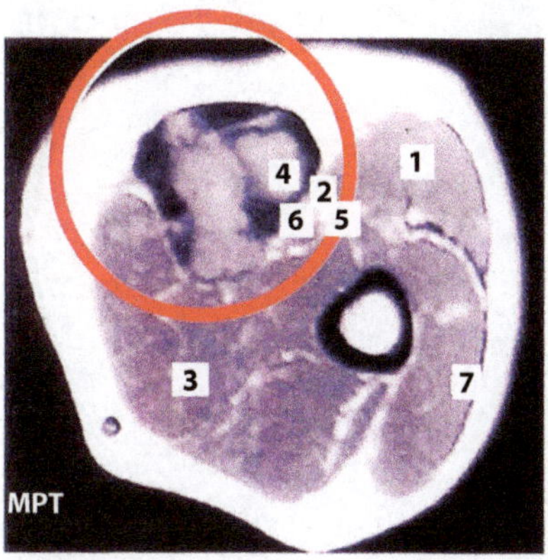

Figure 140 Pathotopography of the transverse section of the thigh in the MR-tomogram. 1 – sartorius muscle; 2 – subcutaneous nerve; 3 – external broad muscle; 4 – mesenchymal fibrosarcoma of the thigh; 5 – contents of the leading channel; 6 – the femoral artery and vein; 7 – adductor muscles

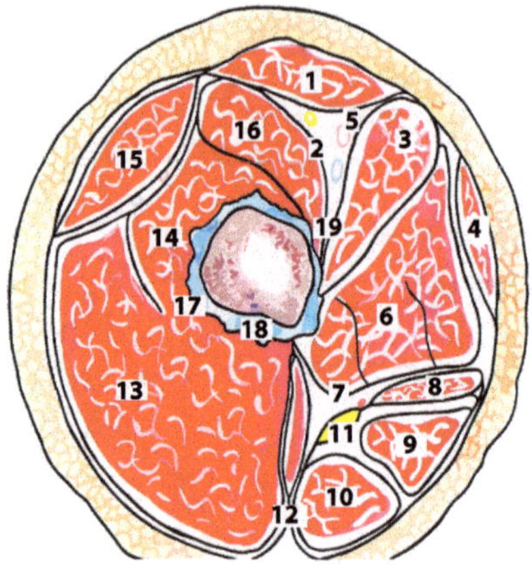

Figure 141 Transverse section of the thigh. Cartilage tumor, cancer "*in situ*".
1 – sartorius muscle; 2 – internal intermuscular septum and femoral vessels; 3 – long adductor muscle; 4 – tender muscle; 5 – perforating artery; 6 – a large adductor muscle; 7 – posterior intermuscular septum; 8 – semimembranous muscle; 9 – semitendinous muscle; 10 – biceps muscle; 11 – sciatic nerve; 12 – external intermuscular septum; 13 – external broad muscle; 14 – medium broad muscle; 15 – straight muscle of the thigh; 16 – internal broad muscle; 17 – cartilaginous tumor; cancer "*in situ*"; 18 – leading channel; 19 – subcutaneous nerve

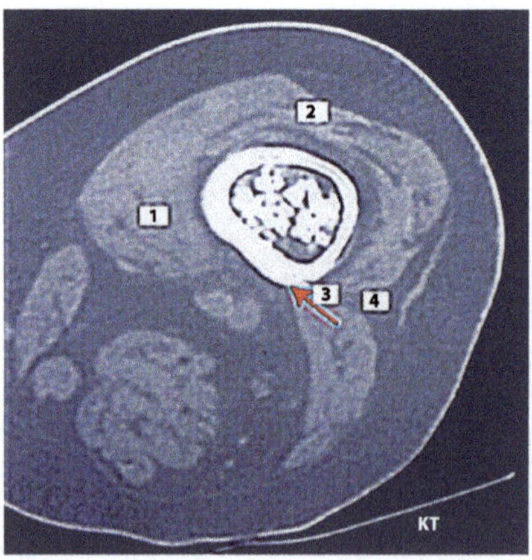

Figure 142 Transverse section of the femur on CT. Cartilage tumor, cancer "*in situ*"
1 – vastus femoris muscles; 2 – rectus femoris muscle; 3 – cartilaginous tumor; cancer "*in situ*"; 4 – adductor canal

When chondromas are formed in the metaphyseal-diaphyseal and metaphyseal-epiphyseal regions of long tubular bones, symptoms are also absent in the initial stages. As the tumor grows, there are passing arthralgias (pain in the joint without signs of inflammation). If the chondroma reaches a sufficiently large size, the development of synovitis is possible.

Increasing in size, the chondroma can sprout a significant portion of the diameter of the bone. At the same time, a strong bone tissue is replaced by a softer and tender cartilaginous, so large tumors, regardless of their location, can cause pathological fractures. The characteristic signs of such fractures are severe pain, crepitation and pathological mobility in the area of injury in the absence of trauma or after minor trauma.

Chondroma, growing from the base of the skull and located in its cavity, early begin to squeeze the brain. The tumor increases slowly, so the symptoms of constriction appear gradually. Possible dizziness, headaches, movement coordination disorders (vestibular ataxia), paralysis and paresis of the limbs.

On a computer tomogram, a cartilaginous tumor is visible, the cancer is "*in situ*", the resulting channel is visualized since it is not compressed by a limited pathological process.

On a computer tomogram, a cartilaginous tumor is visible, the cancer is "*in situ*", the resulting canal is visualized, since it is not compressed by a limited pathological process.

The calcaneal spur is a bone growth in the form of a beak, spine or wedge located in the region of the calcaneus of the calcaneus from its plantar side or in the area of attachment of the Achilles tendon. Symptoms of the calcaneal spur are caused by permanent traumatization of the surrounding tissues with the development of bursitis, periostitis and dystrophic changes. At the beginning of the clinical manifestation of the calcaneal spur, patients complain of periodic pain while walking, then there are sharp initial pains that decrease during walking. To confirm the diagnosis, a radiograph of the foot in the lateral projection is performed.

Pathotopography (Figures 143 and 144). Due to various reasons, excessive tension of the plantar fascia occurs up to microscopic ruptures in the place of its attachment to the calcaneus. The fascia itself consists of a large number of fibers – these are the micro-ruptures that cause inflammation of the fascia and, as a result, acute pain.

A little below, look at the correct image of spur formation, which clearly shows the stages and mechanism of hook formation as a result of increased mechanical impact. Here they are:

1. the tear of the plantar aponeurosis
2. because of the tears of the fasciae of the aneurysm, an inflammatory process arises.
3. the inflammatory process leads to the ossification of the plantar tendon (aponeurosis) at the point of attachment to the heel and the formation of the hook.

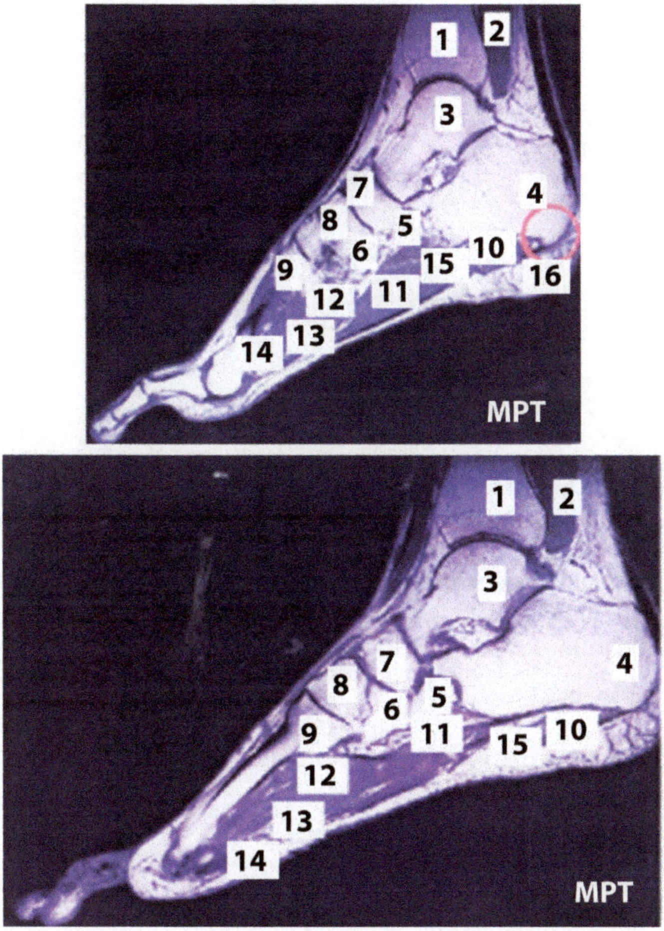

Figure 143 Sagittal section of the foot on the MR-tomogram.
The MR-tomogram shows all the anatomical structures indicated in thecaption to the figure. The followinginteraction of bones is seen: the plus (9), behind it the tarsal bones are thelateral sphenoid bone (8), the scaphoid bone (6), the intermediate sphenoidbone (7), the cuboid bone (5), the talus bone (3), calcaneus (4) In contrast to the normal cut in thisplane, a change in the shape of the calcaneus contour (4) is visualized on the MR-tomogram of the patient with calcaneal spur (16) with the pressure of theunderlying muscle, which removes the little finger (10)

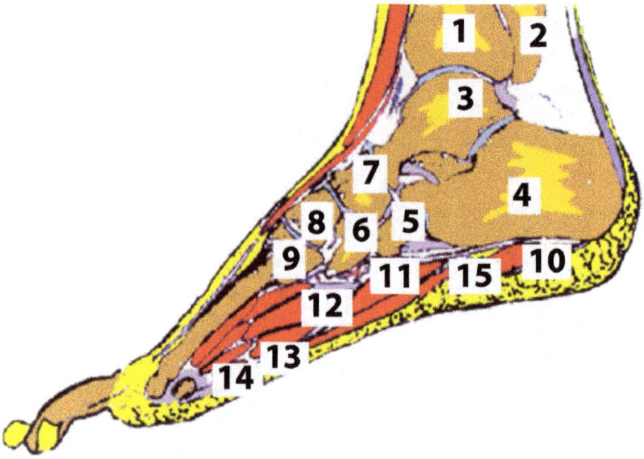

Figure 144 Sagittal section of foot.
1 – os tibia; 2 – os fibula; 3 – talus; 4 – calcaneus; 5 – os cuboideum; 6 – os naviculare; 7 – os cuneiforme intermedium; 8 – os naviculare laterale; 9 – metatarsus; 10 – m. abductor digiti minimi; 11 – m. flexor digitorum brevis; 12 – m. adductor hallucis (caput obliquum); 13 – m. lumbricalae; 14 – m. adductor hallucis (caput transversum); 15 – n.; a. et v. plantaris lateralis; 16 – calcaribus calcaneus

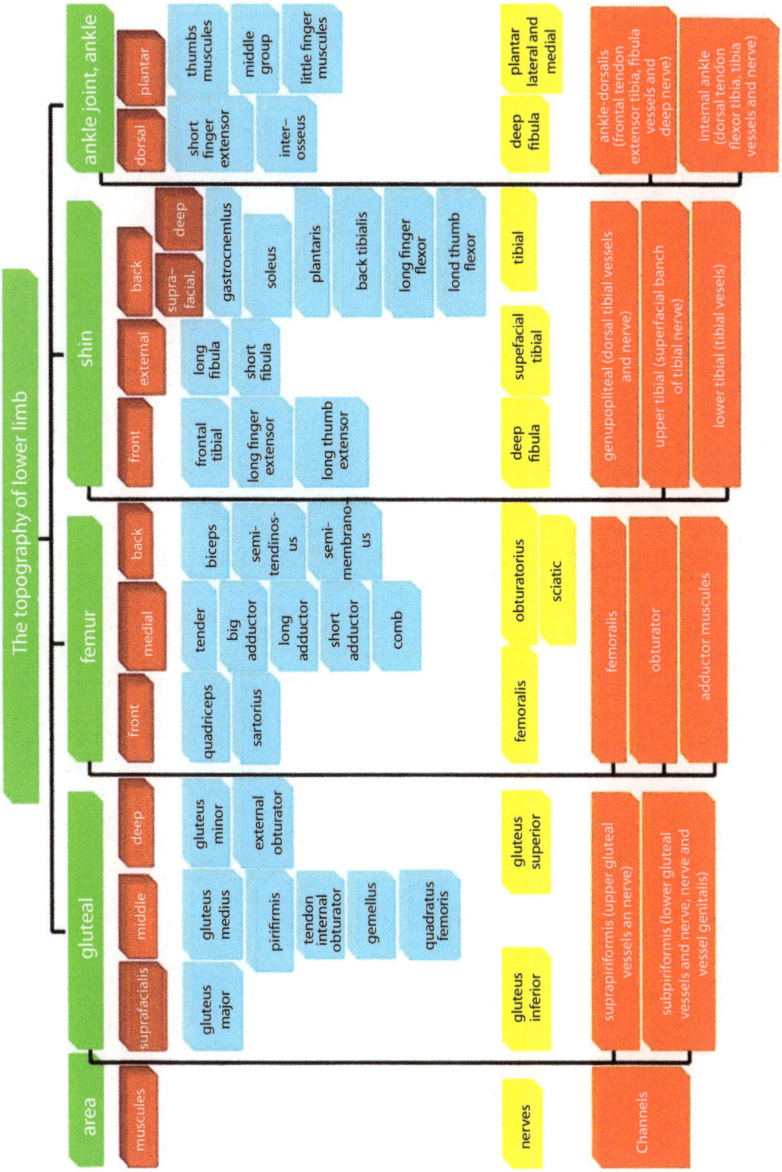
The topography of lower limb
area
muscules
nerves
Channels
gluteal
suprafacialis
middle
deep
gluteus major
gluteus medius
pirifirmis
tendon internal obturator
gemellus
quadratus femoris
gluteus minor
external obturator
gluteus inferior
gluteus superior
suprapiriformis (upper gluteal vessels an nerve)
subpiriformis (lower gluteal vessels and nerve, nerve and vessel genitalis)
femur
front
medial
back
quadriceps
sartorius
tender
big adductor
long adductor
short adductor
comb
biceps
semi-tendinos-us
semi-membrano-us
femoralis
obturatorius
sciatic
femoralis
obturator
adductor muscules
shin
front
external
back
supra-facial.
deep
frontal tibial
long finger extensor
long thumb extensor
long fibula
short fibula
gastrocnemlus
soleus
plantaris
back tibialis
long finger flexor
lond thumb flexor
deep fibula
supefacial tibial
tibial
genupopliteal (dorsal tibial vessels and nerve)
upper tibial (superfacial banch of tibial nerve)
lower tibial (tibial vesels)
ankle joint, ankle
dorsal
plantar
short finger extensor
inter-osseus
thumbs muscules
middle group
little finger muscules
deep fibula
plantar lateral and medial
ankle-dorsalis (frontal tendon extensor tibia, fibula vessels and deep nerve)
internal ankle (dorsal tendon flexor tibia, tibia vessels and nerve)

Conclusion

The color atlas of human topography and pathotopography is simultaneously an educational, scientific and practical illustrated monograph for physicians of all specialties, students of medical schools of all faculties, and the teaching staff of medical schools. Along with the systematic presentation of the entire topography of an adult and a child in our three-volume manual, pathotopography is presented for the first time. Classical topographic anatomy includes layered topography, postmortem and lifelong normal toporaphy, variational, individual and projectional topography, congenital and acquired developmental defects. Pathotopographic anatomy consists of surgical, therapeutic, gynecological, traumatological, oncological topography and etc. It is developed with ultrasound, MRI, CT, transillumination topography. It is practically important to differentiate for diagnostics and treatment of organ and inorganic, functional pathotopography with the determination of the viability of target organs by our original methods of cigolography – pulseomotorography, mucocellular transport. Numerous monographs, patents of inventions and doctoral and candidate dissertations of our students have been devoted to the determination of viability in pathotopography.

Applied aspects of topography and pathotopography are associated with a description of the vascular-neural canals, external and internal hernias, weakness places, localizations and pathways of abscesses and phlegmons. Learning graphs of logical structures include various areas, formations, organs, blood supply, venous outflow, lymph drainage, innervation of organs. The ways of collateral blood supply are presented. Our inventions allow to stop bleeding with the preservation of collateral circulation.

Thus, we hope that the information presented in the atlas will contribute to a more effective diagnostics and treatment of focal and diffuse pathology, the prevention of iatrogenic complications. In the manual, for the first time we laid the foundations of a new discipline of pathotopography with new specific qualitative and quantitative differential-diagnostic signs of osteoarthritis, rheumatoid arthritis, hemarthrosis, benign and malignant thyroid and mammary gland formations. At the same time, the new ways of determining the effectiveness of treatment have been found.

This atlas was created in the name of the patient and for the sake of his doctors treating him. For the same purpose, in our department of topographic anatomy and operative surgery we create the world's only operative surgery museum, where exhibits are presented that combine topography, pathotopography and operative surgery. These exhibits are based on our atlas. Here are some of them.

Figure 145 The museum of operative surgery.

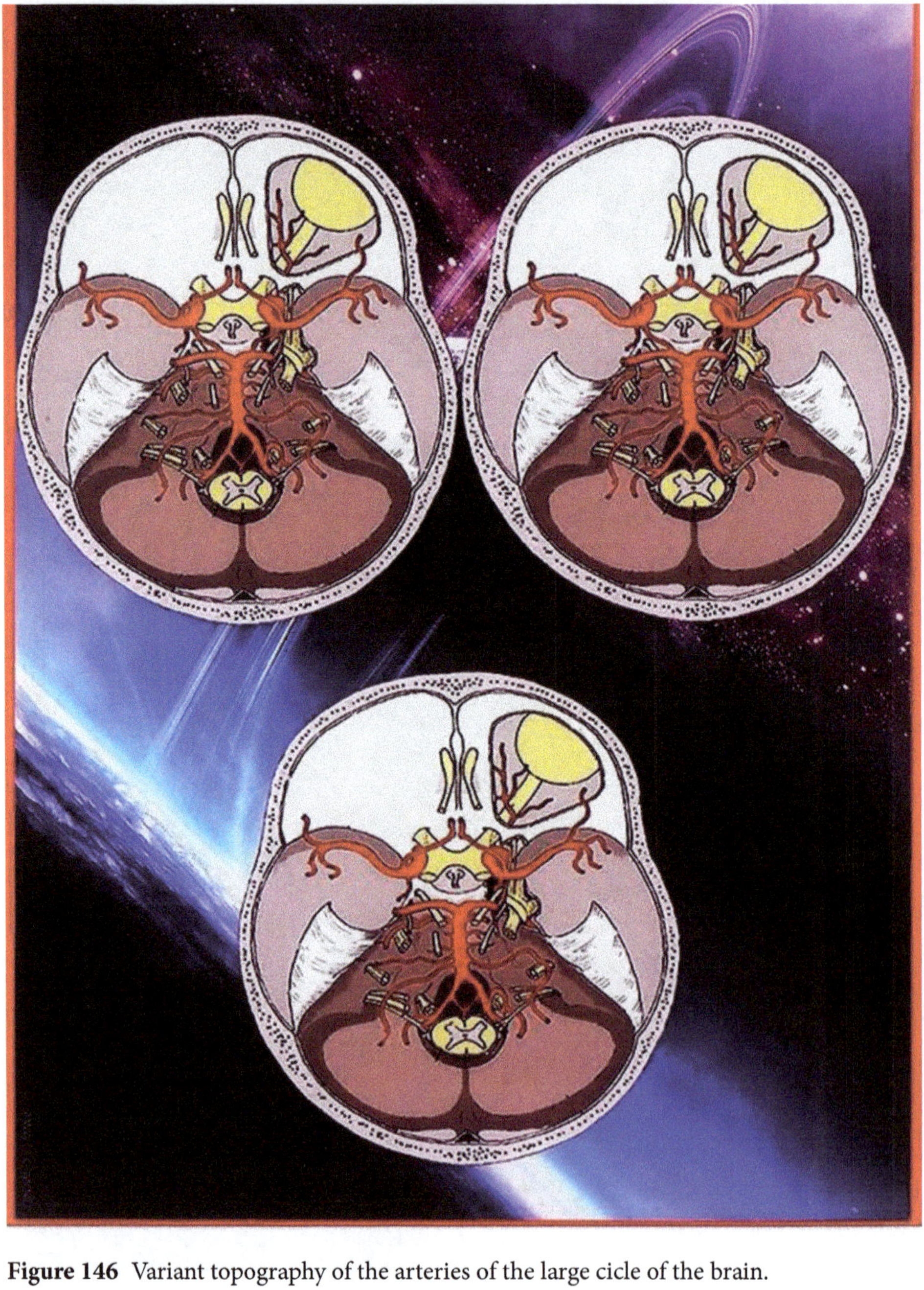

Figure 146 Variant topography of the arteries of the large cicle of the brain.

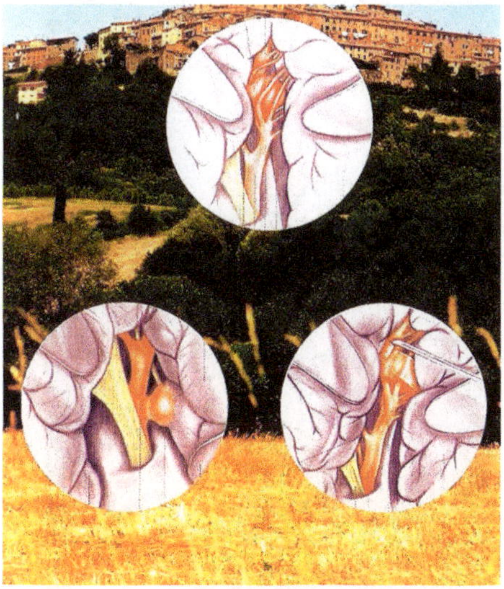

Figure 147 Variant pathotopography of aneurisms of arteries of the large circle of the brain.

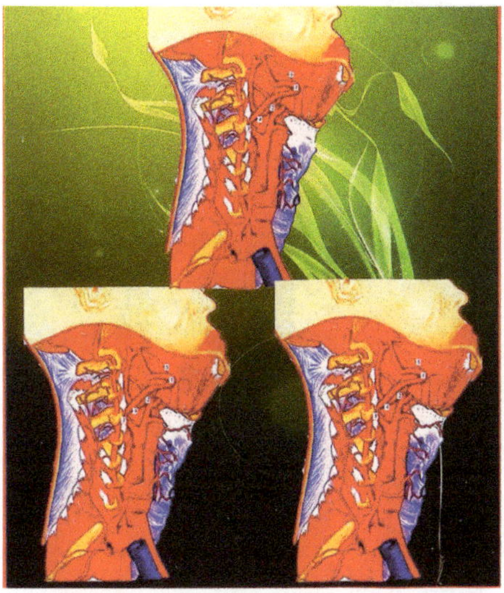

Figure 148 Variant topography of the arteries of thyroid.

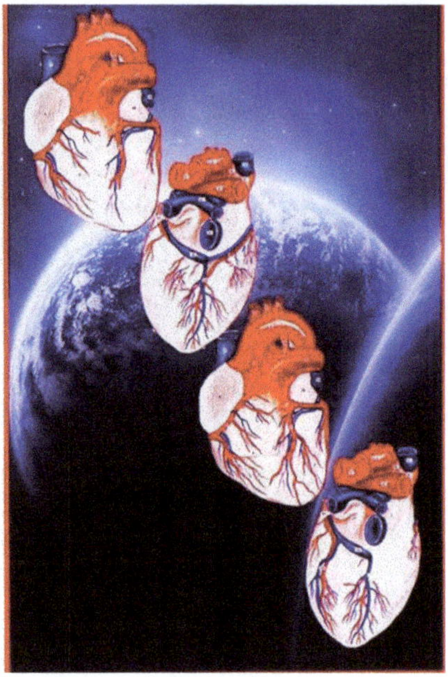

Figure 149 Variant topography of coronary arteries of the heart.

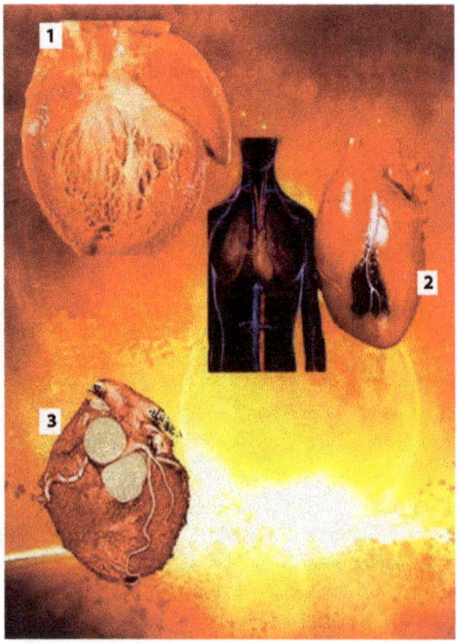

Figure 150 Variants of diseases of coronary arteries of the heart.
1 – cardiosclerosis; 2 – ischemic heart disease (IHD); 3 – arteriosclerosis

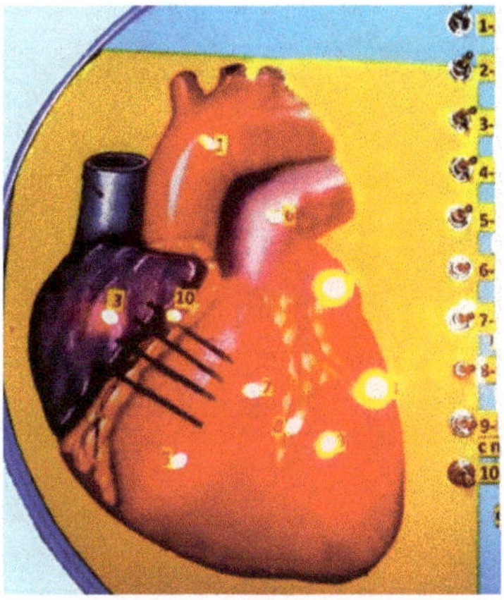

Figure 151 Suturing of wound of the heart.
1 – the aorta; 2 – the left atrium; 3 – the right atrium; 4 – the left ventricle; 5 – the right ventricle; 6 – the pulmonary truncus; 7 – wound of the heart near the coronary arteries; 8 – the coronary artery; 9 – U-shaped sutures with teflon linings; 10 – threads that further will be subjected to formation of nodes

Figure 152 The method of protection intraintestinal anastomoses on vascular pedicle.
1 – transplant of the omentum in the form of ring; 2 – anastomosis; 3 – U-shaped serous-muscular sutures

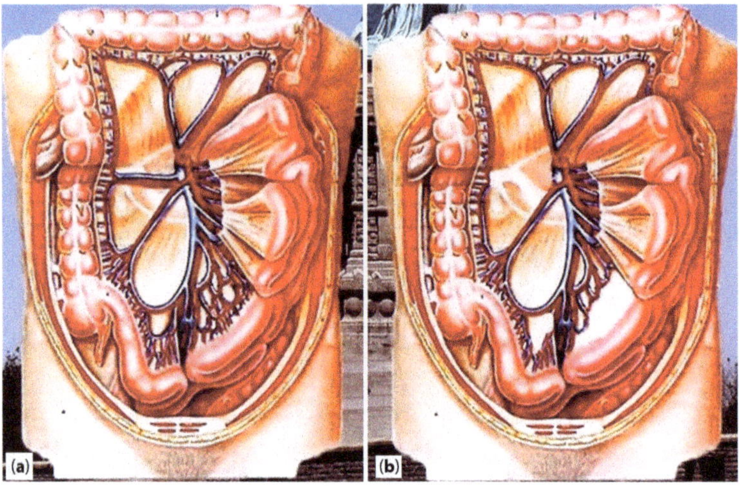

Figure 153 Blood supply of the bowel.
a – blood supply of the ascending colon by a. et v. iliocolicae, a. et v. colicae dextrae.
b – blood supply of the ascending colon by a. et v. iliocjlicae

Figure 154 The Roux resection in the stomach.
1 – the esophagus; 2 – the fundus of the stomach; 3 – the corpus of the stomach; 4 – the small curvature of the stomach; 5 – the large curvature of the stomach; 6 – gastrointestinal anastomosis; 7 – intestinal anastomosis; 8 – stump of duodenum; 9 – transverse colon

Figure 155 The atergrade appendectomy.
1 – ilium; 2 – appendix veriformis; mesoappendix; 3 – colon ascendens; 4 – tenia libera; 5 – gaustrae coli; 6 – appendices epiploicae; 7 – caecum; 8 – ligature on the mesoappendix; 9 – ligature on the appendix

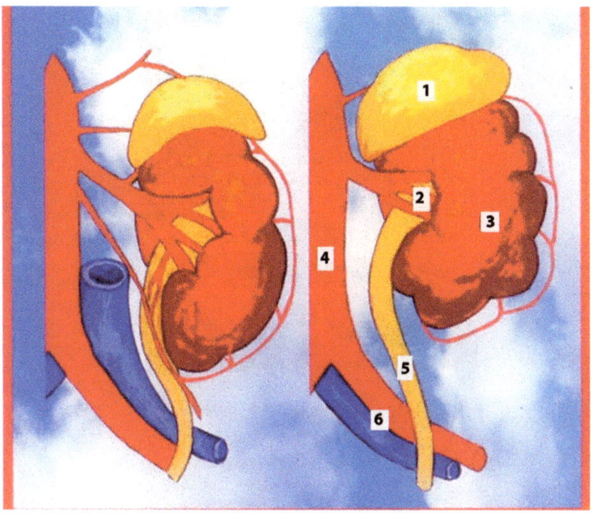

Figure 156 Variant toporaphical anatomy.

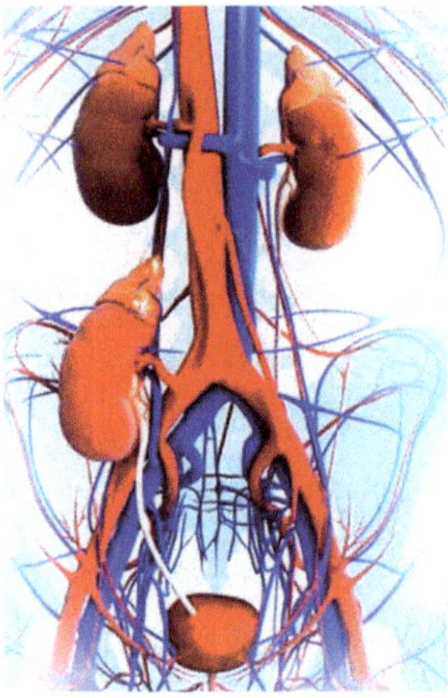

Figure 157 Kidney transplantation.

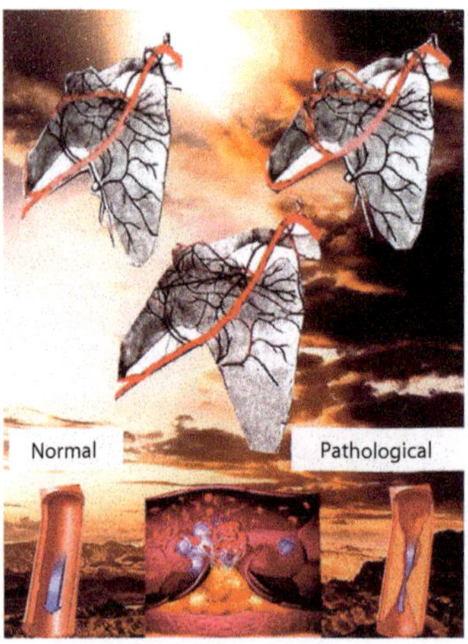

Figure 158 The occlusion of the axillary artery.

Figure 159 The museum of the topographic and pathotoporaphic anatomy. The patents.

Figure 160 The museum stand.

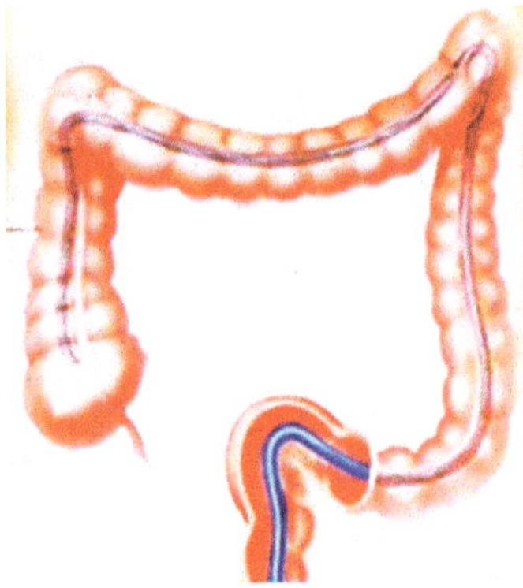

Figure 161 The method of studying viability, motility of hollow organs without surgical intervention (Patent №1463311).
Authors: Z. M. Seagal, S. L. Tochilov.
Patent claims: the method is characterized by the fact that in order to improve the intravenous blood supply of the organ in the postoperative period, the perforation is performed in the form of alternating screw incisions of the right and left directions with a displacement of one relative to the other by 0.5 steps and with shortening of each slot from both sides by 0.1 steps.

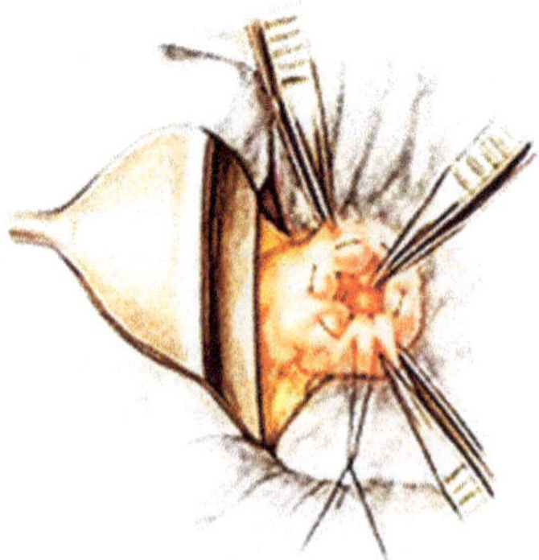

Figure 162 The method of preventing the inconsistency of intestinal stumps (Patent №2189601).
Authors: Z. M. Seagal, E. V. Halimov.
Patent claims: it differs in that the intestine is tied, stepping 0.5–1 cm from the junction of the contralateral intramural vessels, the suture stitch is placed in the junction zone of these vessels and the thread of the pouch is hold over the visible intramural vessels.

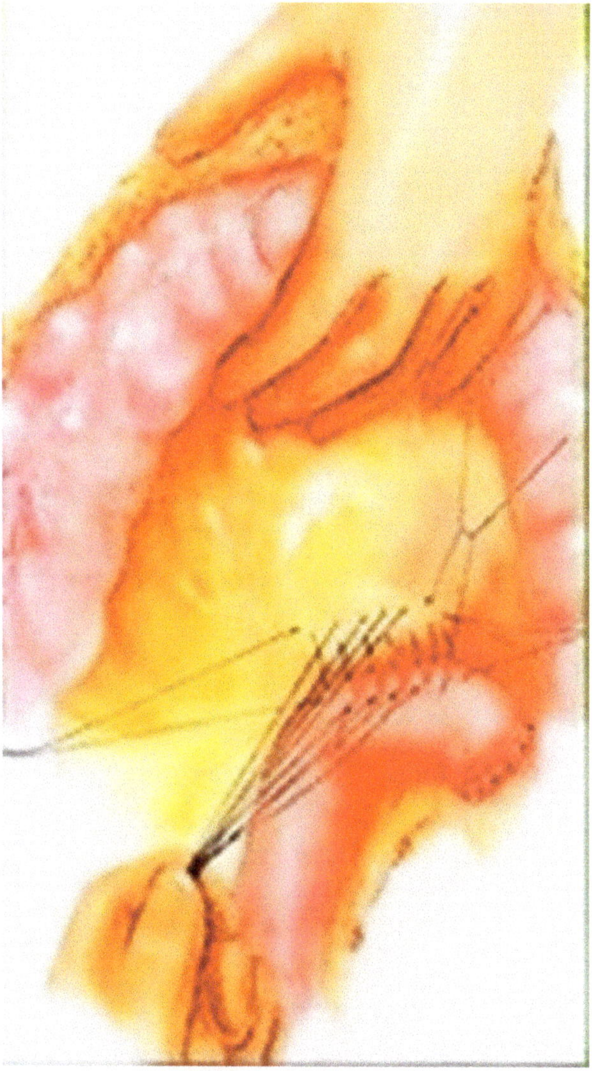

Figure 163 The method of protection of interintestinal anastomoses (Patent №2008105301).
Authors: Z. M. Seagal and others.
Patent claims: the method includes stitching of anastomises to the great omentum. It differs in that the transplant of the omentum on a vascular pedicle with an intraorganic pulse all along is cut out, the suture stripe is wrapped around the transplant in the form of a ring and fixed with U-shaped serous-muscular sutures to the wall of the intestine on both sides along the the course of large intramural vessels and the great omentum is sied to the mesentery of the intestine in the avascular zone.

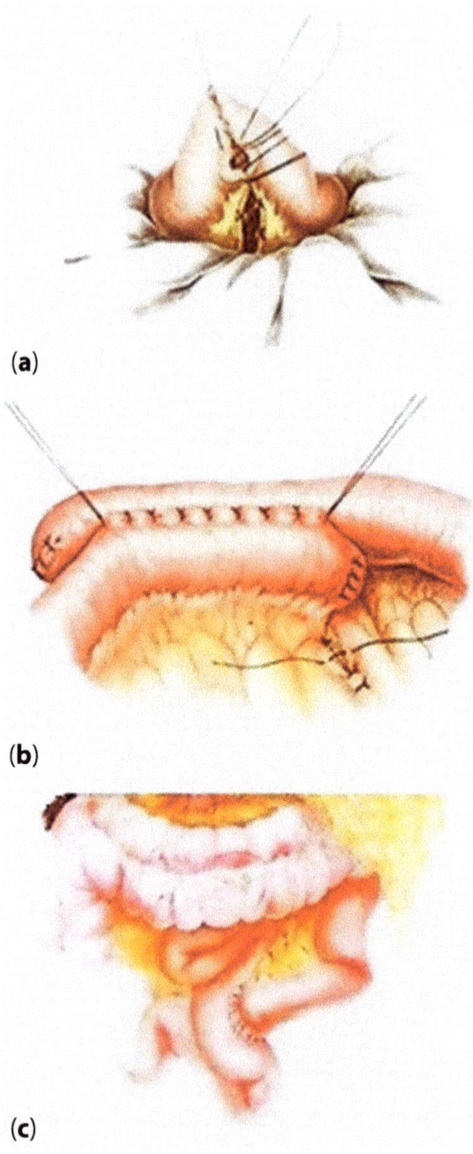

Figure 164 The interintestinal anastomoses.
a – "end-to-end"; b – "side to side"; c – "end to side"

About the Author

Z. M. Seagal is an Honored Scientist of the Russian Federation and the Udmurt Republic, Honorary Academician of the Izhevsk State Medical Academy, Head of the Department of Operative Surgery and Topographic Anatomy, Doctor of Medical Sciences, and Professor.

Topographical and Pathotopographical Medical Atlas of the Pelvis, Spine, and Limbs

The atlas presents topographical and pathotopographic human anatomy for both the adult and the child. The section "Pelvis" includes topographic anatomy of the pelvis both in men and women. Layer topographic anatomy, variant, computer, and MRI-topographic anatomy are presented in this section. The surgical anatomy of congenital malformations of the pelvis includes diverticulum of the bladder, bladder exstrophy, and atresia of rectum and anus. In the section "Spine", nerve roots, individual, and age differences of the spinal cord and topography of the spine are described. The section "Limbs" includes the topographic and pathotopographic anatomy of upper and lower limbs. All the pictures are colorful and original. The atlas is written in accordance with the educational program of medical universities of the Russian Federation. The original graphs of logical structures are presented according to the sections of topography and congenital malformations. This allows an effective study of the subject.

This atlas is intended for the students of General Medicine, Pediatrics and Dentistry faculties, as well as for interns, residents, postgraduate students, and surgeons. The monograph is intended to be used by physicians, junior physicians, medical residents, lecturers in medicine, and medical students.

Also of Interest

By the same author

Ultrasonic Topographical and Pathotopographical Anatomy: A Color Atlas, by Z. M. Seagal and O. V. Surnina, ISBN 9781119223573. Using ultrasonic technology to create full color detailed pictures, this atlas details the topographical and pathotopographical anatomy of the human body, as a useful reference for medical professionals and students alike. *NOW AVAILABLE!*

Check out these other titles from Scrivener Publishing

Compendium of Biophysics, by Andrey B. Rubin, ISBN 9781119160250. The most thorough coverage of biophysics available in a handy, easy-to-read volume, perfect as a reference for experienced engineers or as a textbook for the novice. *NOW AVAILABLE!*

Fundamentals of Biophysics, by Andrey B. Rubin, ISBN 9781118842454. The most up-to-date and thorough textbook on the fundamentals of biophysics for the student, professor, or engineer. *NOW AVAILABLE!*

Ethics in the University, by James G. Speight, ISBN 9781118872130. Examining the potential for unethical behavior by all academic staff, both professionals and nonprofessionals, this groundbreaking new study uses documented examples to show where the matter could have been halted before it became an ethics issue and how to navigate the maze of today's sometimes confusing ethical academic arena. *NOW AVAILABLE!*

Reactive Oxygen Species: Signaling Between Hierarchical Levels in Plants, by Franz-Josef Schmitt and Suleyman I. Allakhverdiev, ISBN 9781119184881. Not available anywhere else, this groundbreaking volume presents a novel and unique approach to understand the complex chemical network behind the internal network structure of plants, a valuable tool for students of physics, biological physics, chemistry, and biology, but also for sociologists and economists as well as for scientists in any field. *NOW AVAILABLE!*

Lightning Source UK Ltd.
Milton Keynes UK
UKHW020559050719
345627UK00003B/4/P

9 781119 614258